T0255810

Number Story

Number Story

From Counting to Cryptography

PETER M. HIGGINS

<u>C</u>

COPERNICUS BOOKS

An Imprint of Springer Science+Business Media

Peter M. Higgins, BA, BSc, PhD
Department of Mathematical Sciences, University of Essex
Wivenhoe Park, Colchester, UK

An imprint of Springer Science+Business Media, LLC

Mathematics Subject Classification (2000): 11-01
British Library Cataloguing in Publication Data
A catalogue record for this book is available from the British Library

ISBN 978-1-4471-6851-5   ISBN 978-1-84800-001-8(eBook)
DOI 10.1007/978-1-84800-001-8

Printed on acid-free paper.

© Springer-Verlag London Limited 2008
Softcover re-print of the Hardcover 1st edition 2008

9 8 7 6 5 4 3 2 1

Springer Science+Business Media

springer.com

Preface   ix

chapter 1   **The First Numbers**   1
How Should We Think About Numbers?   5
The Structure of Numbers   8

chapter 2   **Discovering Numbers**   17
Counting and Its Consequences   23

chapter 3   **Some Number Tricks**   31
What Was the Domino?   34
Casting Out Nines   35
Divisibility Tests   39
Magical Arrays   49
Other Magic Number Arrays   57

chapter 4   **Some Tricky Numbers**   61
Catalan Numbers   65
Fibonacci Numbers   67
Stirling and Bell Numbers   72
Hailstone Numbers   75
The Primes   77
Lucky Numbers   84

chapter 5   **Some Useful Numbers**   85
Percentages, Ratios, and Odds   85
Scientific Notation   88
Meaning of Means   90

chapter 6	On the Trail of New Numbers	101
	Pluses and Minuses	104
	Fractions and Rationals	105
chapter 7	Glimpses of Infinity	117
	The Hilbert Hotel	120
	Cantor's Comparisons	122
	Structure of the Number Line	128
	Infinity Plus One	133
chapter 8	Applications of Number: Chance	137
	Some Examples	141
	Some Collectable Problems on Chance	148
chapter 9	The Complex History of the Imaginary	165
	Algebra and Its History	168
	Solution of the Cubic	174
chapter 10	From Imaginary to Complex	185
	The Imaginary World Is Entered	189
	The Polar System	195
	Gaussian Integers	198
	Glimpses of Further Consequences	200
chapter 11	The Number Line under the Microscope	209
	Return to Egypt	212
	Coin Problems, Sums, and Differences	216

Fibonacci and Fractions   221
Cantor's Middle Third Set   225

chapter 12 Application of Number: Codes
           and Public Key Cryptography   229
Examples from History   230
Unbreakable Codes   238
New Codes for a New World of Coding   242
Simultaneous Key Creation   244
Opening the Trapdoor: Public Key Encryption   251
Alice and Bob Vanquish Eve with Modular
Arithmetic   255

chapter 13 For Connoisseurs   263
Chapter 1   263
Chapter 3   268
Chapter 4   271
Chapter 5   281
Chapter 6   283
Chapter 7   289
Chapter 8   296
Chapter 9   300
Chapter 10   303
Chapter 11   309
Chapter 12   312

Further Reading   315

Index   319

Preface

Numbers are unique, there is nothing like them and this book reveals something of their mysterious nature. Numbers are familiar to everyone and are our mainstay when we feel the need to bring order to chaos. In our own minds they epitomize measured rationality and are the key tool for expressing it. However, do they really exist? They certainly don't exist the way cats and football teams exist, or even the way colors and feelings exist, but more in the way that words exist. Words have meanings and the meaning of a number, what the number 'is', is about overall matchings that allow us to measure and compare things that might otherwise have little in common, such as the value of oil, of a taxi cab, and of the services of its driver.

And collectively numbers represent the one thing in the world that is free and inexhaustible. It is therefore natural to try and understand them as much as we can.

The opening chapters of this book will re-acquaint the reader with numbers, both seen as individuals and taken all together. Throughout the first four chapters, we generally stick to discussing ordinary, whole counting numbers. The fifth chapter looks at some practical issues surrounding number use that, by involving arithmetic operations, lead us out of an environment where everything is given in solid, discrete chunks.

Chapter 6 explains how it is that through carrying out the standard operations on numbers, we discover new number types, including the irrational. In the subsequent chapter we visit infinite collections and see how they can be compared to one another and how the set of real numbers as we call them knit together to form the number line, something we examine with a mathematical magnifying glass later in the book.

The historical development of Number History is, like all history, a complex thing but one that seems to have resolved itself to the extent that number systems now enjoy agreed status among mathematicians and certainly form a central pillar of our understanding of the world. Throughout the text we inform the reader of various historical snippets associated with the evolution of the subject and a little about individual number pioneers. This culminates in Chapters 9 and 10 where we summarize the development that took place in Europe during the formative period from the 16th to the end of the 19th centuries.

And we do look at direct applications of numbers, most notably in Chapter 8, which is all about chance, and again in Chapter 12 that concerns itself with the clandestine world of codes and secret ciphers, which have proved the major new field of applications of pure number ideas.

The book is written to be read straight through by any interested reader although dipping and browsing might be equally rewarding. We do however provide one final chapter, *For Connoisseurs*, in which some of the particular claims and examples in the text are worked through in mathematical language for the benefit of those readers who would appreciate complete explanation. An asterisk in the text indicates that more is said on the topic in the notes of the final chapter. This is the only chapter of the book that makes free use of mathematical notation and ideas. The level of difficulty here varies as determined by the nature of the material in question but all readers will be able to glean something from examining some of the notes at the end of the book. Finally there is a short closing section giving direction to other fine books and Web sites for you to enjoy.

I hope this little book will allow my readers to grasp something of a very big story, the Story of Numbers.

Colchester, England, 2007   Peter M Higgins

chapter 1

The First Numbers

'All is number', said Pythagoras over 2,500 years ago. By this he meant that, at its deepest level, reality is mathematical in nature and could be expressed in terms of numbers and the ratios between them. Was he right? The short answer is no, as he himself is said to have discovered.

It is true that the disciples of Pythagoras revealed how aspects of the world were governed by number. Pythagoras is best known for his celebrated theorem that explains how the lengths of the sides of a right-angled triangle are related to one another. The modern interpretation of this is that the exact distance between two points can be found from their co-ordinates. This discovery provided a tool allowing the precise calculation of spatial separation from other measurements and so represented a real breakthrough. More surprisingly perhaps, Pythagoras is said to have discovered that pure musical harmony is determined by simple ratios. Flushed with success, it must have seemed to the Pythagoreans that any aspect of the world would yield to analysis through number, for these were astonishing revelations. The clarity and simplicity offered by the laws of Pythagoras was of a kind never previously encountered.

It came therefore as a shock when Pythagoras found that numbers themselves rebelled against his rule, for he is credited with also discovering that certain lengths constructed in his geometry were impossible to express as simple fractions the way his philosophy demanded. In particular, he found that you cannot measure the diagonal of a square with the same units with which you measure the sides. However fine you make the scale, the tip of your diagonal will always lie between two of your scale marks. This is due to the fundamental nature of numbers, and has nothing to do with limitations on the accuracy of your ruler or the sharpness of your vision. It is a mathematical fact of life. What might be dismissed by us however as an annoying curiousity was viewed as a catastrophe by the Pythagoreans, for it undermined their whole outlook by which they sought to explain nature through simple number ratios. Even from these early classical times then, there were problems with the view that everything could be reduced to numbers.

Despite their limitations however, numbers have not retreated but rather crowd into our lives relentlessly. As far back as the early 17th century, Galileo advocated as a guiding principle that we should measure everything we can and learn to measure those things we cannot. Embracing this philosophy has yielded rich results and in calling for a measurement we are being asked to produce a number.

There is however a natural resentment provoked when this seems to be taken too far. Attempts to call upon numbers as a tool for understanding music and poetry often meet with scorn. The very idea spoils the magic and it is natural to sneer at the possiblity and hope for failure. In this it still seems that we are on safe ground as numbers rapidly begin to lose authority in the artistic realms. To be sure, music has a mathematical side to it, as Pythagoras discovered, and that aspect is well worth understanding. However, a purely analytical approach to the arts yields pretty thin results. Good music is not produced by calculations, and the more this avenue is explored the poorer the offerings produced.

Mistakes along these lines are in any case far from new. Right throughout history and across cultures we can find examples where numerical ideas are introduced in a misguided way that eventually leads to nothing of interest. To simply assert, for example, that even numbers are female and odd numbers male, or the reverse, is not helpful. Artificial attempts to make up the laws of nature have never worked and say more about the human mind than they do about the real world: simple ideas designed to appeal to our fancy may be comforting and even fun, but are rarely true.

As a backlash to the constant call for numbers and percentages, there is an agressive tendency in the arts today to reject anything to do with systematic or scientific thinking. This is a frame of mind that some great artists, Leonardo da Vinci for one, would have found puzzling. I wonder if this yearning to be released from the straitjacket of logical thinking is more born of frustration, stemming from a lack of creativity, which is blamed on the way numbers have taken over our lives. Constantly measuring things seems to be the very opposite of spontaneity, leading to a dislike of numbers that are seen as a tiresome and inhibiting burden. Perhaps the very way we think has become enslaved by the rule of numbers that acts as a limitation on us all, retarding freedom of thought and spirit.

Let me assure you nonetheless that numbers are not evil but rather are naturally interesting. The problems we may have with them, and the destructive uses they may be put to, are of our own making. It is best on the one hand to appreciate that there are going to be limitations to their legitimate uses but, on the other, admit that it is not always easy to tell in advance where those limitations will lie. One surprising facet of numbers is the odd way they have of invading other branches of math and science, quite out of the blue. For example, until around 30 years ago no-one had any idea that the so-called trapdoor functions on which our internet security codes are based would come about through ideas about ordinary numbers, but more of that part of the story later.

Galileo (1564–1642) was right in his belief in the value of measurement[1]—perhaps we should however add the modern caveat that we should resist the temptation to pretend that we have measured something when we have not. How often, for instance, do we hear in modern life an expert say that he is 90% sure of an outcome—not 92% or 88%, but 90%. The figure lacks true meaning if there is no way of calculating it. However, we often feel obliged to produce a number even when we do not have one so we can fall into the trap of simply making them up in order to sound more authoritative. In the absence of real information, a vague statement may be correct and a precise one with a number in it merely a form of wishful thinking made in order to sound more informed and convincing in the face of uncertainty.

Most times when we meet up with numbers, we are called on to interpret them in a particular context, which might be about money, people, or the pressure of a gas. However, the subject of this book is the numbers themselves and how our understanding

---

1 Although a relatively minor figure, Nicholas of Cusa (1401–1464) had advocated two centuries earlier that knowledge must be based on measurement.

of them continues to evolve. It is only right that we begin by examining the kind of thoughts we have when we come across these mysterious things called numbers.

How Should We Think About Numbers?

When we mention a particular number, let us say for example, sixteen, all of us have a mental picture of the two numerals 16. This is somewhat unfair to the number in question as we are immediately stereotyping sixteen as $10 + 6$. Why should we think of sixteen as $10 + 6$ when it could equally well be described as $9 + 7$ or, more symmetrically as $8 + 8$? This habit, of course, comes from our unswerving use of the number ten as the base of our number system: our expression of a number implicitly displays it as a sum of powers of the number ten. For instance, when we write 2008 we mean $2 \times 1000 + 0 \times 100 + 0 \times 10 + 8 \times 1$. As you may know, we would be equally entitled to use another base such as twelve for our number system and different civilizations of the past did indeed use different bases: the Mayans sometimes used twenty, the Babylonians employed base sixty, while modern computing systems are based on two or small powers of two such as four, eight, and sometimes even base sixteen, which is known as *hexadecimal*. Since $16 \times 16 = 256$ we can cover that many possibilities with two symbols in base 16 (although we need to introduce new individual symbols for the six numbers normally denoted by 10, 11, 12, 13, 14, and 15). Two hexadecimal digits are all you need to represent any number in the range from 0 to 255 inclusive, a common spread used, for example, to specify colors. As we shall see in a later chapter, comparison of numbers in different bases can also be used in subtle ways to reveal the nature of how numbers order themselves into a line.

We shall say more about this in due course but we should first ask the more fundamental question: Why do we introduce a base at all when we want to deal with numbers? You might think that there is no way of coping with number matters without referring to some base or other. However, we do just that more often than you may realize in everyday life. Suppose for example we have a childrens' party where we want to give every child a toy. All that matters is that there are at least as many toys as children and we can check this without counting: we could simply write each child's name on a toy and as long as we don't run out of toys before we have exhausted all their names, no-one will go away disappointed. In doing this we establish that the number of toys is at least as great as the number of children and we do it without counting up either collection. We do not need to know how many children or how many toys we have in order to show that the number of toys is sufficient. We therefore have solved this problem about numbers without introducing base ten or any other base to do our calculation. This example also serves to show that number is very much about pairing members of one set with another, a very important idea.

Use of a particular base does allow us however to express numbers in an efficient and uniform manner that makes it easy to compare one number to another and to perform the arithmetical operations that arise through counting. A base of a number system is akin to placing a grid scale on a map. It is not intrinsic to the object but is rather like a system of co-ordinates imposed on top as an instrument of control. Our choice of base is arbitrary in character and the exclusive use of base ten saddles us all with a blinkered view of the set of counting numbers, 1, 2, $\cdots$. Only by lifting this veil can we see numbers face-to-face for what they truly are.

Various local number systems cropped up in many cultures, but all exploited the grouping of collections into equal size lots, often of size ten. The efficacy of a base in your arithmetic only comes into its own once you introduce the *positional principle* in representing numbers where the value of a numeral depends on its place within the number string. No ancient society, not even the sophisticated Greeks, developed a complete positional numbering system such as we have where the value of a numeral depends on its position within the number and full use is made of a zero symbol to indicate that a certain power of the base is absent (recall our example of 2008). It was in the early centuries of the first millenium that such a complete numbering system came into being in India, with a symbol for 0 called *sunya*, which is the Hindi word for empty. It passed to Europe via the Arabs so that our number system is known as Hindu–Arabic.

Not having a proper positional approach to arithmetic is a real handicap for most practical purposes. Yet not being mentally trapped in a base ten world did make it easier and more natural to study numbers in their own right. The freedom the Ancients enjoyed by default we may reclaim for ourselves simply by shedding the base ten mantle for a time and thinking of numbers in terms of the intrinsic properties they may or may not enjoy.

Having emancipated ourselves in this way, we see that it is more natural to focus on the special factorization properties of a number as these correspond to appealing geometric displays. The number sixteen for example is a perfect square, naturally represented by a four-by-four square of dots, and since four is itself a square we notice that sixteen is a perfect fourth power as it is equal to

$2^4 = 2 \times 2 \times 2 \times 2$. In fact sixteen is the first number after 1 that is a perfect fourth power, making it very special indeed. This is a reason why it is often used as a base itself in computing systems, as opposed to base ten, which is the traditional base we use for the accidental reason that we have ten digits on our hands.

If we suspend the habit of thinking of numbers simply as servants of the science of measurement, and take a little time to study them without reference to anything else, much is revealed that otherwise would remain hidden. The natures of individual numbers can manifest themselves in ordered patterns in nature, like the spiral head of a sunflower, (which represents a so-called Fibonacci Number), and so are worthy of a thorough investigation in their own right. Simple questions about numbers, such as how they may be written as the sum of squares, have led to mathematical structures of great beauty and intricacy. Instinctively mathematicians will follow signposts of that kind as they often lead to very unexpected destinations that would not be stumbled upon in any other way.

For convenience I shall still write the individual numbers that I call your attention to in the usual way in base ten but we will not be emphasising that representation: rather we shall regard it more as a name for the number we are presently thinking about.

The Structure of Numbers

One of the glories of numbers is a fact so self-evident that it may easily be overlooked—they are all different. Each number has its own structure, its own character if you like and the personality of individual numbers is important. Take the number six. Six is a product of two smaller numbers, namely two and three, and so forms what we might call a *rectangular number*: one that can be represented as a rectangular array of dots. A number n that can be written as a product of two smaller numbers, $n = a \times b$ say, can be drawn as an $a \times b$ rectangle of dots. (We normally save time and space by writing the product $a \times b$ of a pair of unspecified numbers, a and b, simply as ab.) Rectangular numbers are more often called *composite numbers* as they are composed of smaller factors. Numbers that are not rectangular in this way are known as *primes*. Prime numbers such as 2, 7, and 101 cannot be displayed as a proper rectangle but rather only as a single line of dots. In words, a number is prime if it *cannot* be written as the product of two smaller factors. (A definition that precludes 1 from joining the list of primes: the first prime is 2.) The primes are structurally important as they form the multiplicative building blocks from which all numbers can be put together: for example 60 is a composite number that is a product of prime numbers: $60 = 2 \times 2 \times 3 \times 5$. Any composite number can be broken down into a product of factors which, if not themselves prime, can be broken down further until we recover the *prime factorization* of our number. It turns out that this factorization is unique—there is only one way to factor a number as a product of primes. However you attack the factorization of your number, if you keep factoring its factors you will always end up with the same collection of prime factors. This is a crucial property of numbers that is exploited in diverse applications of the subject from coding to logic. Indeed perhaps the greatest unsolved problem in all mathematics is the Riemann Conjecture, which is intimately connected with this so called Fundamental Theorem of Arithmetic that says that the prime factorization of a number is unique.*

It is hard to over emphasise the importance of the uniqueness of prime factorization. Reading this, you may wonder at the fuss. To be sure, if prime factorization were not unique, everyone would have heard about it by now. True as that is, the following example shows that it is not the kind of thing that can be taken for granted. Consider the numbers in the sequence, 1, 5, 9, 13, 17, 21, $\cdots$: that is the numbers of the form $1 + 4n$, as n takes on the successive values 0, 1, 2, 3, 4, 5, $\cdots$. This collection of numbers forms a multiplicative number system in its own right in that if we multiply any two numbers from this sequence together, we remain within the sequence: for example $9 \times 17 = 153 = 1 + (4 \times 38)$. Some numbers, like 153, can be factorized into a product of other numbers in the set of numbers formed by the sequence. Some however cannot, in which case let us call the number *primal*. Ordinary primes in the sequence, such as 5 and 13 are primal, as is 9, as 9 cannot be factorized within the set ($9 = 3 \times 3$ but 3 is not in our set).

It is clear that any number in this sequence can be broken down into a product of primal numbers: we argue just as with primes for either the given number is already primal, or it is not, in which case it can be broken into smaller factors from the set that we break down further until this can be done no more and we are left with a product of primal numbers. However, primal factorization is not always unique: $693 = 21 \times 33 = 9 \times 77$, which gives two different primal factorizations of $693 = 1 + (4 \times 173)$.

The moral of the story is that uniqueness of prime factorization is special, and, although familiar, is not self-evident for here we have a similar number system in which it does not apply.

Returning to our featured number 6, we note that the property of being rectangular is hardly a remarkable one. However 6 is also a *triangular number*: since 6 = 1 + 2 + 3 it can naturally be regarded as a triangular array of six dots with one in the first row, two in the second, and three in the third. The previous triangular number is 3 = 1 + 2 and the next is 10 = 1 + 2 + 3 + 4. We usually allow 1 to be admitted among the list of triangular numbers as well so that the first five of them are 1, 3, 6, 10, and 15. The 10 and 15 triangles can respectively be seen in the pin array of 10-pin bowling and the 15-ball rack of red balls in snooker. Triangular numbers form a more exclusive set than the class of the very common rectangular numbers.

The number 6 is also what we might call a choice number: the number of ways of choosing a pair from a group of four children numbers six in all. If the children are Alex, Bart, Caroline, and Daniel the six pairs we may form can be listed as AB, AC, AD, BC, BD, and CD, where we are paying no regard to the order in which we list the children within a pair, meaning for example that we regard AB and BA as representing the same pair. It turns out that any triangular number is also a choice number in a similar way as the nth triangular number is also the number of ways of choosing a pair from a family of $n + 1$ objects. Again we shall explain this further in Chapter 4.

The fact that 6 = 1 + 2 + 3 has another interpretation that occurs much more rarely in the infinity of the number system as this sum shows that 6 is the sum of all its smaller factors. The Pythagoreans called such numbers *perfect*. One should always be wary of a seductive name but on this occasion it is not misplaced: for a number to be the *sum* of its factors in this way does suggest it has a special internal balance and it is one that is indeed very rare. The next four perfect numbers are 28, 496, 8128, and 33,550,336. A lot is known about the even perfect numbers but, to this day, no-one has been able to answer the basic question of the Ancients as to whether there are infinitely many of these special numbers, although there is a correspondence between them and a particular class of prime numbers. What is more, no-one has found an odd one, nor proved that there can be no odd perfect number. Will we ever find out?

Finally 6 has a truly unique property in that it is both the sum and product of all of its smaller factors: $6 = 1 \times 2 \times 3 = 1 + 2 + 3$ and it is also the sum and product of a sequence of consecutive numbers. There is certainly no other number like this. Indeed it is often easy enough to find peculiar properties of small numbers that characterize them—for instance 3 is the only number that is the sum of all the previous numbers while 2 is the only even prime (making it the oddest prime of all).

The nth triangular number arises from summing all the numbers from 1 up to n together. If we replace addition by multiplication in this idea we get what are known as the *factorial* numbers. The first factorial is 1, the second is $2 \times 1 = 2$, and the third, as we have already seen, is $3 \times 2 \times 1 = 6$. Factorials come up constantly in counting and enumeration problems such as the chances of being dealt a certain type of hand in a card game like poker. For that reason they have their own notation: the nth factorial is denoted by $n! = n \times (n-1) \times \cdots \times 2 \times 1$. The triangular numbers grow reasonably quickly, at about half the rate of the squares, but the factorials grow much faster and soon pass into the millions and millions: for example $10! = 3,628,800$. The exclamation mark, a notation introduced by Christian Krempe in 1808, alerts us to this rather alarming rate of growth.

It is fair to say that small numbers tend to be more special than larger ones—the closer a number is to the beginning of the number line, the more likely it is to display some genuinely unique trait. This however is only a rule of thumb and some large and very large numbers turn out to be intrinsically special. The number 12 is an *abundant* number meaning that it is exceeded by the sum of the factors less than itself: $1 + 2 + 3 + 4 + 6 = 16$. It is rare for an odd number to be abundant and no small odd number is. However it is possible and the first example turns out to be 945. Readers might care to check for themselves that when we sum all the factors of 945 the result is the larger number 975. It is possible, if you know a bit about these things, to see this coming: $945 = 3^3 \times 5 \times 7$, a standard formula then gives that the sum of the factors, *including the original number*, will then be given by $(1 + 3 + 9 + 27)(1 + 5)(1 + 7)$ from which, upon subtracting 945, the figure of 975 results.*

Mathematicians who are intimately connected with number theory can get to know individual numbers so well that they become old friends. A famous conversation between Hardy and Ramanujan concerned the number 1729 of a taxi cab. When Hardy carelessly suggested the number was dull, the little Indian genius immediately disabused him, pointing out that 1729 was the smallest number that was the sum of two cubes in two distinct ways: $1729 = 1^3 + 12^3 = 9^3 + 10^3$.

There are numbers that are especially annoying such as 561. It behaves a lot like a prime number without being one. A basic property of a prime number p that is particularly important in coding theory is that it satisfies the Fermat Lemma which says that for any number a, a^p leaves the same remainder as does a when divided by p. For example, if we take the prime $p = 5$ and put $a = 8$ we can check that both the numbers 8 and $8^5 = 32,768$ leave the remainder 3 when divided by 5. However this is not generally the case for composite numbers p: for example if we replace the prime 5 by the composite number $p = 4$ and put $a = 7$ we see that the remainders when 7 and $7^4 = 2401$ are divided by 4 are respectively 3 and 1 and so are not the same. It would be convenient if this property provided a test for whether or not a number p were prime but it does not. The composite numbers p that always pass this test are called the *Carmichael Numbers* and $561 = 3 \times 11 \times 17$ is the smallest of them. These numbers are rare but, coincidentally perhaps, Ramanujan's number, 1729, turns out to be another one, as is 2821. In the year 1992 it was proved nevertheless by Alford, Granville, and Pomerance that, as with the primes, Carmichael numbers continue without end so there is no way past them.

Primes are elusive in a way that some other types of numbers are not. If we want, for example, a very large square, we just write down a big number and multiply that number by itself and there we have it. However, although it has been known since before the time of Euclid (300BC) that there are infinitely many primes*, they are not so easy to generate and it seems that we need to go out hunting for them. We cannot manufacture primes the way we can with the squares—we are limited to testing one odd number after another, although there are various tricks that facilitate the endless search. On the one hand no-one has proved that it is impossible to find a way of readily generating primes at will, but on the other hand, no-one can can claim to have yet succeeded in doing so.

Primes are common enough among the first few thousand numbers but they slowly become rarer and rarer as we move into the realm of the very large. This is not surprising as a large number has potentially more possible factors than a small one. At any time in the history of mathematics, there is a largest known prime number. The current champion has over four million digits and would take a month just to write down in ordinary base 10 notation. It can however be written as one less than a power of two: $2^{13,466,917} - 1$. Since there are always larger prime numbers waiting in the wings to be discovered, the pre-eminent status of this number is but a passing thing.[2]

However an example of an extraordinarily large number with a special status that can never be lost is

$$8080\ 17424\ 79451\ 28758\ 86459\ 90496\ 17107$$
$$57005\ 75436\ 80000\ 00000.$$

This is the size of the so-called *Monster sporadic group*. A little explanation is in order. A *group* can be thought of for our purposes as the collection of all symmetries of an object: movements such as reflections and rotations that leave a patterned object such as a square or wallpaper design looking as it did before. Mathematical groups are a topic that only emerged in the early 19th century from the study of the solutions of certain equations involving powers of orders higher than two. However they have proved strikingly pervasive, penetrating almost all of mathematics and physics: crystallography and coding are but two fields where they arise. The short explanation for this is that they give an algebraic hold on the geometric notion of symmetry, allowing us to perform calculations based around that idea.

Mathematics always searches for ways in which complicated objects are made up of smaller and simpler parts. A *simple group*

---

2 Indeed it has passed during the preparation of this book: at the time of writing the largest known prime is the 44th so called Mersenne prime, $2^{32,582,657} - 1$ found in 2006. The record is being broken regularly at present thanks to the international GIMPS project that has enlisted tens of thousands of enthusiasts working with their computers searching in parallel. See http://primes.utm.edu/largest.html.

is to groups what a prime number is to numbers, in that a simple group cannot be built from smaller groups, in a sense that can be made precise, but need not concern us here. There are four main sources of simple groups but, in addition to these types there are exactly 26 so-called sporadic simple groups that lie outside of the mainstream. It is now known that there are no more than these 26 exceptional groups. They are simple in the technical sense only and generally are enormous in size and complexity. The Monster is the largest of them all and was constructed in 1982 by Robert Greiss as a group of rotations of 196,883-dimensional space. The size of the Monster is the 54-digit number given above. That number is therefore special and will remain special for all time. It is a permanent feature of the mathematical landscape. The extent of its significance will only be revealed as years go by and the full story of numbers unfolds.

chapter 2

Discovering Numbers

Despite their familiarity, it should be appreciated that numbers have no physical existence but rather are abstractions elicited from the real world. Two sets are said to have the same number if the members of the sets can be paired off, one against the other, as in Seven Brides for Seven Brothers. The number of one finite set is less than that of the other if the first can be paired off with just a portion of the second, as in our example where we gave toys to the children at the party. This gives the set of counting numbers a natural ascending order. Since we all have been taught to count from childhood it is not easy to appreciate what a difficult idea counting represents. It must have been hard to realise and put into words that a pair of rabbits and a couple of days are instances of the same thing. The practical upshot of course is that the man with the rabbits has one meal for each of the next two days.

Once we have grasped the notion of number it is natural to give names to the first few of them: one, two, three, four etc are the ones we use. If we did not go beyond this stage the process would be little different than that whereby we recite the letters of the alphabet in a particular order. The contexts are not entirely parallel however: the first twenty-six numbers have the natural order mentioned above whereas the order of the letters of the alphabet is quite arbitrary: although the *names* of our numbers could be anything we fancy, the natural ordering of the numbers is intrinsic and is not something of our making. It is the arbitrary nature of the order that we impose on the alphabet that accounts for the effort children are called on to make so as to remember the order in which letters appear in the dictionary.

What is adequate for the alphabet however is not good enough for numbers as the first set is finite—we reach the end after inventing twenty-six names, while the collection of numbers is infinite and stretches away indefinitely. What is more, in practice we need to make use of lots of numbers—any civilization will need to be able to count into the hundreds and thousands on occasion so there is a call to devise some kind of number identification that goes beyond the naive approach of creating an ever-growing list of different words for different numbers.

We can mitigate against this difficulty a little by agreeing that certain numbers are represented by a single symbol: for instance in Roman numerals X and V stand for ten and five respectively. However the fundamental problem would still remain, that being that it is impractical, indeed impossible to have a single unique symbol for every number. Sooner or later we are forced to make use of the *Addition Principle* whereby some numbers are represented as the sum of two smaller ones. For instance, in Roman numerals there is no special symbol for fifteen—we just write XV to indicate the number that results from taking a group of ten and adjoining to it a group of five.

It would seem that the discovery of the Addition Principle is a very natural one for we see it put to use in all the ancient civilizations of the Middle East, Europe, and Asia. Additions based on ten were also prevalent. As mentioned before, the ancient Bablyonians made use of both base twelve and base sixty from which come the worldwide practices of dividing the day into twenty-four hours and the full circle into 360 degrees. Another remnant of base sixty is in French where there are no new names for numbers past 60 up till 100: 70 is soixante-dix (60 and 10), 80 is quatre-vingt (four twenties), 90 is quatre-vingt-dix etc. Belgian French speakers however grew tired of this and introduced the new names septante, octante, nonante etc for these numbers. Most number systems however took up the option of grouping into tens, which allowed for the recording of fairly large numbers through use of a short string of symbols. Unfortunately the Just Good Enough Principle was generally adopted: once a way of writing numbers was invented that was adequate for day-to-day business it became completely entrenched and no effort seemed to have been made to improve further and certainly not to replace it with one that was better.

Even the mathematically sophisticated Greeks did not take basic arithmetic seriously enough to break free of a quite primitive notation. One explanation for this is that matters of accounting were considered the province of mere slaves and quite unworthy of higher study. Whatever the reason, the pattern of reckoning of the Greeks was little more advanced than in other ancient cultures. (Indeed the Babylonian system was fundamentally superior, as will be explained.) It could well have been that ancient accountants had a host of practical tricks for doing their sums—certainly they made good use of simple devices such as the abacus (counting board) and no doubt had their own idiosyncratic methods of mental arithmetic that were communicated to the next generation by word of mouth and through example. That part of the History of Mathematics is largely lost with only accidental glimpses being available to the modern scholar.

The Greeks represented the numbers 1–9 by the first nine letters of their alphabet, and used a similar string of symbols for the multiples of ten from 10–90, while a further set of nine stood for each of the numbers 100 through to 900. For example, λ and β stood respectively for 30 and for 2 so that the number 32 was written as $\lambda\beta$. At first glance this may look as efficient as our notation but it is not. The Addition Principle is being exploited but no real use is being made of position. If we swap the digits of 32 we get the different number 23. However that does not apply to $\beta\lambda$, which could still only mean $2 + 30 = 32$. The Greek version of 23 would have been $\kappa\gamma$, as κ stood for 20, while γ was the third letter of the Greek alphabet and so could stand for 3. In this way all numbers up to one thousand can be recorded by strings of length no more than three. In the early days of the system, that might have proved fairly adequate. Before too long though, it became necessary to deal with numbers going into the thousands. Rather than start from scratch, the old system was modified in an ad hoc fashion in order to cope. It became understood that putting a comma before a symbol meant that symbol was to be multiplied by 1000 so that, for example $,a$ was the representation of 1000. This must have proved good enough for practical purposes.

There were sporadic attempts to do better. In the third century AD the Greek mathematician Diophantus went one step further in using a dot to indicate that the preceding number was multiplied by a myriad (10,000). He gave the example $,a\tau\lambda a. \ ,\varepsilon\sigma\iota\delta$ which we accordingly translate as 13,315,214 as the number $1000 + 300 + 30 + 1$ represented by the first group of four symbols is multiplied by ten thousand because that quartet is followed by a dot, while the latter four stand for 5000 + 200 + 10 + 4 in turn. In this way we see that it is not too difficult to adapt what might appear a clumsy system to write down numbers running into the millions. Indeed Archimedes in the 3rd century BC could boast in his book the *Sand Reckoner* that he could represent a number greater than the number of grains of sand required to fill the universe (at least the universe of the Greek World).

We might still object that this way of representing numbers would not lend to pen and paper arithmetic. However that is a very modern objection as the ancient world did not have cheap paper. Difficult sums were performed on counting frames so their method of writing numbers only had to be good enough to record the answers and the ingredients that made them up. Number notation did not need to go far beyond a shorthand for writing out numbers in words, and so it never did.

The origin of the system of Roman numerals is very obscure but was probably Etruscan, which was a civilization that pre-dated the Romans on what is now the Italian penisula. Roman numerals were indeed used by the Romans and persisted right through medieval times and survive, mainly for decorative purposes, in modern European culture. In addition to the symbols for one, five and ten mentioned above were also symbols for fifty, one hundred, five hundred and one thousand, which were respectively L, C, D, and M. That a film was made in 2003 is indicated at the end of the credits by the Roman numerals MMIII, while the year 1673 was written MDCLXXIII. Similarly to the Greek system, the Romans embellished their number symbols to indicate multiplication by a large power of ten. For example two hundred thousand and one million could be indicated by placing boxes around the symbols II and X respectively to show these quantities were to be increased by a factor of one hundred thousand.

In the Greek system the meaning of a symbol was fixed and did not depend on where that symbol occurred in the string. The Roman system however did make limited use of position in that the *Subtraction Principle* was sometimes employed. To explain: the Roman strings in our two previous examples were conventionally written in descending order of size, as in the Greek system. However, in Roman numerals we can write a minor symbol before a major one, such as in IV, and the meaning in such circumstances is that the value of the minor symbol is *subtracted* from the major. This gives an alternative to writing four simply as IIII. Similarly the numbers nine and forty can be written respectively as IX and XL. It seems though that the Romans themselves made little use of the subtraction principle that only came into consistent use in Europe after the invention of the printing press.

Although the Roman system was capable of improvement, the direction that was being followed was a dead end. A practical modern arithmetic required that these ancient systems be completely scrapped and replaced with a positionally based notation for numbers. The Babylonian base sixty system is remarkable in that it did make use of position: their cuneiform symbol for 1 could mean 1, or 60, or 60×60 depending on its position within the string of symbols. Unfortunately this great idea was not exploited to its full potential because of the lack of a zero symbol to act as a placeholder, although occasionally it seems they did leave an empty space. The empty register was only used however for intermediate places and never in the terminal position, as we do for instance when we write a number like 70. Overall then their notation could still be confusing through its ambiguity.

To treat 0 as a number in the same way as the positive integers 1, 2, etc is an enormous psychological hurdle. In order to realize the

full potential of arithmetic and algebra, numbers other than positive counting numbers need to be embraced because arithmetical operations lead us out of the domain of the natural numbers
and into the realm of other number types. As long as we tie our
mathematics to some particular interpretations of it we are always
liable to be bogged down in irrelevancies. Even today many people
are unhappy with the idea of negative numbers whilst imaginary
numbers that arise from the square root of the negatives are considered completely beyond the pale, requiring some special kind
of mind to comprehend. None of this is true, but acknowledging
this attitude allows us to have some sympathy with the enduring
prejudice against using 0 in the same way as other numbers. Full
use of a zero symbol may have just been regarded as ugly and
unnecessary. This is still with us today and can be seen in the use of
computer read forms. Often we are told to use two digits for a date
so that if you were born on the 8th of February 1964 for example
you should record your date of birth as 02/08/64 (or in Britain
as 08/02/64). Some of us baulk at pandering to the uniformity
demanded by the computer system and refuse to place a 0 symbol
in front of a number and so just write 2/8/64. Despite being a
trivial thing, people can get very upset and stubborn about these
matters.

Counting and Its Consequences

It seems that counting is a natural development for humans as
the discovery of counting can often follow the need for it. There
is a recent example of tribal peoples of the Arctic who lacked a
system of counting but quickly developed one of their own upon

encountering western civilization that had the effect of surrounding them with objects that needed to be tallied up.[3] If we are to count up to a fair sized number it is natural to break the task into two or more stages, from which point it is but a short jump to the idea of addition. The pause in the counting process corresponds to the idea of generalized counting where we do not always begin the count at the number 1. If we have two collections of stock, each of which has been counted separately, then in performing the full count we are doing an addition sum. Pure counting itself can then be viewed as a special kind of addition where at each stage the number to be added is just 1.

And so we see that addition arises from counting and addition then leads on to the other three arithmetic operations of subtraction, multiplication, and division. Historically it is less clear which of subtraction or multiplication was the next to develop. Intuitively subtraction might seem the most natural, being addition's opposite and indeed it is usually the second operation introduced to school children.[4] Subtraction takes away or undoes an addition (which may in reality never have been performed) to leave you with fewer objects than you began with. Although subtraction looks simple enough, from the mathematical viewpoint it is very awkward. It does not behave as nicely as addition—when you perform two subtractions in succession it matters where you put the brackets,[5] something you never have to worry about when adding up. What

---

3 See Stephen Pinker's, *The Blank Slate*.

4 The oldest book in which the familiar + and − signs appear in print is a commercial arithmetic *Rechenung auff allen Kauffmanschafft* by Johann Widman of Leipzig published in 1489. The equals sign is later and an English innovation due to Robert Recorde in 1557.

5 $(8 − 4) − 2 = 4 − 2 = 2$ but $8 − (4 − 2) = 8 − 2 = 6$: subtraction is not *associative*.

is more, you can't carry out some subtractions for it seems that you cannot subtract a larger number from a smaller one, whereas you may add any two numbers that you like. This is somewhat disconcerting, and even children can be left with the feeling that we are somehow missing a trick. When they object, the explanation is usually along the lines that if we have three ducks on a pond, then we can't take away four ducks because there aren't enough ducks. This is still unsatisfactory because the natural symmetry of the addition–subtraction operations has broken down and in order to justify this we have ended up talking about ducks. Although children may not express their doubts in this way, the thought may still linger at the back of their minds that things are not quite right and that there is more to be said.

Multiplication on the other hand throws up no such difficulties as it is a special form of repeated addition: 4×3 means $4 + 4 + 4$. The one query that does arise is, why is it worthy of attention? This may be a surprising question if it has not occurred to you before, as multiplication is so familiar. The answer however lies in experience, which has shown that this particular kind of addition arises constantly in real problems: for instance finding the area of a rectangular field.

Much of mathematics has arisen by seeking to take a fruitful idea one step further. Repeated addition of the same number leads to multiplication so perhaps repeated multiplication of the same number, the corresponding special kind of multiplication, is also an important notion.

If we replace the plus sign by a multiplication sign in the previous sum we get $4 \times 4 \times 4$, which we normally write as 4^3, and indeed this type of repeated multiplication is important: in this case the answer represents the volume of a cube of side length 4.

This operation of raising to a power is called *exponentiation*.[6] Not all aspects of the pattern persist however. Exponentiation is not *commutative*: $3^4 = 81 \neq 64 = 4^3$ so, unlike multiplication and addition, the order in which we introduce the numbers into the operation matters.

There is a mathematical operation based on repeated exponentiation. An alternative notation for 4^3, especially used on computer keyboards is the uparrow notation: $4 \uparrow 3$. The $\uparrow\uparrow$ then acts as follows: $4 \uparrow\uparrow 3$ means $4 \uparrow 4 \uparrow 4$ which equals 4^{4^4}, so that 3 is the length of the tower of 4's. If we continue with this idea of replacing the operation by repetition of the previous one, we generate a series of extraordinarily huge numbers the likes of which had never been expressed until the 20th century.

The numbers $1 \uparrow 1$, $2 \uparrow\uparrow 2$, $3 \uparrow\uparrow\uparrow 3 = 3 \uparrow\uparrow 3 \uparrow\uparrow 3$, etc are called the *Ackermann Numbers*. The first Ackermann number is $1 \uparrow 1 = 1$, while the second is $2 \uparrow\uparrow 2 = 2^2 = 4$. The third Ackermann number is a tower of exponentiated 3's the height of which is itself $3^{3^3} = 7,625,597,484,987$. The size of the fourth Ackermann number, formed by slipping four arrows between two fours, is beyond anything that could be regarded as comprehensible by the mind of a human, and beyond that lie all of the rest. Even 'small' Ackermann numbers have more digits in their decimal expansion that particles in the universe and the nòt so small ones would use up any number of universes just writing them down in the normal way. To give you some idea: the fourth Ackermann number is 4 raised to the power of a tower of 4's, the length of that tower itself

---

6 The exponential notation first appears in *Triparty en la science des nombres* by Nicolas Chuquet around 1500. He made use of positive, negative, and zero powers applied to an unknown quantity. Interpreting roots as fractional powers was practiced as early as the 14th century by Nicole Oresme of Paris.

being 4 raised to the power of 4^{4^4}, which is 4 raised to the power of a number with 155 digits.

The Ackermann numbers are not simply a way of naming particular numbers of incomprehensible size but are used in theoretical computer science to construct examples of calculations which, although possible in principle, could not be carried out in fewer steps than that of the Ackermann numbers. Anything leading in that direction would surely be forever beyond possibility you would think. Richard Conway and Richard Guy in their aptly named *Book of Numbers* introduce what they called as a chained arrow notation that can be used to define numbers that leave the Ackermann numbers in the shade. There may be no end to these notations that allow you to specify numbers that are otherwise beyond calling. In doing this we are following the lead of pioneers like Diophantus, or indeed the ancient Babylonian scribes, who invented ways of writing down numbers of a size that exceeded anything they could possibly ever have use for.

The fourth and by far the most problematic of the arithmetic operations is division. It is not only modern folk who are intimidated by this one: being able to do 'complicated long division sums' was the achievement of only the most intelligent of T.S. Eliot's felines (see *MaCavity, the Mystery Cat* in *Old Possum's Book of Practical Cats*). Division is the reverse operation of multiplication and throws up the same pattern of difficulties as does subtraction: when we write down a series of divisions it matters where you put the brackets (unlike multiplication) so you have to be careful what you mean. Most divisions cannot be carried out in their entirety but leave a remainder. In order to go beyond this point we have to introduce a new kind of number, the fraction. There has never been too much resistance to the use of fractions however as at least some objects can meaningfully be divided into fractional parts. This contrasts with the negative numbers that arise through subtraction which people have always fretted about. The arithmetic of fractions itself is quite complicated nonetheless and if we insist on working with base ten even in the fractional parts we quickly pass into decimal expansions many of which never terminate but go on forever.

Performing subtractions of one number from another is no harder than the corresponding addition. In contrast, long division is a more difficult task than the corresponding long multiplication. To do multiplication you first need to know your multiplications by heart as far as the ten times table, the number you are using as your base. It is then easy to multiply any number by any of the numbers from 1 to 9. Since multiplying by your base number ten just involves slipping a 0 on the end, this reduces any multiplication to summing the results of a sequence of these basic multiplications. This is how the familiar method of long multiplication works.

An example of a division is $3000 \div 18 = 166$ with 12 left over. The number 18 is the *divisor*, and the subject of the division, 3000 in this case, is rather confusingly called the *dividend*. The answer itself has two parts—in this example the *quotient* is 166 while the *remainder*, which is always less than the divisor and may be zero, is 12. Since multiplication is a special type of addition its inverse procedure, division, involves subtraction. More precisely we subtract the largest multiple of the divisor possible from the dividend to leave the remainder. Finding this quotient involves repeated multiplications and subtractions. At each step the effect of the standard long division method is to subtract from what remains of the dividend the largest multiple possible of a power of ten times the divisor. This allows us to build the value of the quotient from

	quotients	remainders
2)	87	
	43	1
	21	1
	10	1
	5	0
	2	1
	1	0
	0	1

Figure 2.1. Calculation to show that 87 in binary is 1010111.

left to right. The final complication comes when the divisor exceeds the dividend and the result is less than 1. Since long division is set up in base ten, such sums are done in decimal format. The pattern of calculations involved is identical to any other long division—we simply need to take care as to the correct placement of the decimal point. In principle though the method of long division is simple-minded: we just keep taking away multiples of the divisor from the dividend for as long as possible and count up how many times this can be done.

Repeated division is the basis of changing a number expressed in one base to another. The remainders at each stage form the digits in increasing powers of the new base. For example, any number can be written as a sum of powers of 2 and this representation is unique. This statement just says that every number can be written in base 2 and that different binary displays always represent different numbers. (Of course this holds for any base, not just for 2 and 10.) For example, to change 87 to binary the calculation runs as given on the previous page (see Fig. 2.1).

We see therefore that 87 in binary is 1010111_2 (the subscripted 2 is there to remind us of what base the number is in). To go back the other way is easy as we just have to add the powers of 2 indicated by the presence of 1's in the display: the first three 1's on the right indicate the presence of 1, of 2^1 and of 2^2 respectively while the first 0 from the right indicates no contribution from the power 2^3—the complete calculation is $1010111_2 = 1 + 2 + 4 + 16 + 64 = 87$. If all this is news to you, you might like to try an example on your own: show in the same way that, in binary, 108 is given by 1101100_2 and convert this binary number back to its base ten representation to check that you have it right.

chapter 3

Some Number Tricks

The science fiction writer Isaac Asimov has recently returned to prominence through the film *I Robot*, which is based on his three Laws of Robotics—basic invioable commandments meant to govern the behavior of machines in order to ensure they did no harm. However in one of his lesser known short stories he contemplated the idea of an 'advanced society', totally dependent on machines, that had forgotten all it ever knew about arithmetic. One day an intrepid soul rediscovers the secrets of how to do sums all by himself and amazes everyone with his powers that are beyond anything they thought possible. Arithmetic suddenly becomes fashionable and the good citizens indulge in feverish speculation as to how to exploit their newly found skill and independence.

Let us hope it is not coming to that but we have to concede that this story sounds much less preposterous today than it did when it was written some forty years ago. Calculators are all very well but we must not allow them to rob us of our intellectual dignity—these devices should be a convenience and not a necessity. What is more, it is not just a matter of pride. A calculator is of limited use to someone who has no real feeling as to the way numbers behave. A mistake in punching in the information will go undetected and the ridiculous answer that results will be accepted unquestioningly. The user needs to know what to expect from a calculator in order for it to be a practical tool.

For this ideal to be maintained we have to ensure that our arithmetical muscles do not atrophy. As a lesson in keeping your numerical freedom alive, I suggest you take the time to master the divisibility tests explained later in this chapter and other arithmetic diagnostics. To decide whether a number is divisible by a given number up to 16 is not difficult and nor is it hard to see why the tests work. Not only will the tests give you a measure of freedom from your calculator but they also, as you will see, let you perform calculations completely beyond the capability of the flashiest hand held machines.

We shall begin with some party pieces. A common mathematical trick exploits what is known as an *algebraic identity*, an equation that is true *for all* values of a number *n* rather than just for one or two solutions. The idea is to make the listener work through a fairly long list of arithmetical operations with his chosen secret number *n*. Unbeknowns to him, the answer is independent of the value of *n* and therein lies the trick.

A magician's trick recently exposed on television used the same principle. The magician let the subject choose from a series of four pairs of face-down cards to complete a 5-card poker hand while the magician's hand was formed from those cards that were left behind. The magician always won because, he claimed, he used subtle psychological tricks to ensure the subject always chose a losing hand. In fact this was a smokescreen—the dealing of the initial card and the set up of the pairs was such that the magician won in every combination. The outcome was independent of the choices being made and all the talk of psychology was a diversion. The magician did exploit psychology but not in the way he would have you believe!

As a novel example of one of these 'mind reading' tricks, try this one yourself. Choose a single digit number, multiply it by nine, and if the answer has two digits, add them together. Subtract five from what you have, giving you a number. Turn the number into a letter by the rule $A = 1$, $B = 2$ and so on. Think of a country beginning with your letter. Finally, take the last letter of your country and think of an animal starting with that letter.

It's odds on that you have Kangaroos in Denmark. A few other outcomes are possible, such as Danish Koalas, and Cats in the Dominican Republic are also consistent with the game. The reason this works is that there are, as it happens, very few countries beginning with the letter $D = 4$—the reason why the arithmetic always ends in four is due to the pattern of the nine times table, as I am sure you will appreciate upon re-reading the instructions in the light of this suggestion.

A more subtle trick involves the listener choosing some secret numerical object (an example will follow in a moment) and then performing several obscure operations. She then reveals the outcome from which the original choice can be discerned immediately although all trace of it seems to have vanished.

What Was the Domino?

Your friend is asked to choose a domino, in effect, a pair of numbers, a and b, with values from 0 (corresponding to a blank) up to 6. You tell her to take one of the two numbers, the choice is up to her, multiply by 5, add 3, double it, add to that the second number on her domino and then tell you the outcome. You then mentally subtract 6 from the number she has given you to yield a two-digit number that is guaranteed to be ab, so that you announce that her chosen domino is a b.

For example, if she chose the 4 5 domino and decided to use 4 as her first number she would, following your instructions, calculate: $4 \rightarrow 20 \rightarrow 23 \rightarrow 46 \rightarrow 51$; you subtract 6 from 51 to give 45 and announce '4 and 5!' If on the other hand she had selected the 5 instead of the 4 to work with she would compute the string of numbers $5 \rightarrow 25 \rightarrow 28 \rightarrow 56 \rightarrow 60$; you would take 6 from 60 and announce her domino as '5 and 4'. Let's try another: blank and 1—deciding to work with the blank she calculates $0 \rightarrow 0 \rightarrow 3 \rightarrow 6 \rightarrow 7$; you then subtract 6 leaving you with 1, which you interpret as 0 1, and so announce correctly 'blank and 1'.

This all makes you look very clever for two reasons. The calculation creates the impression of mixing up the numbers on the domino in an unpredicatable way so it is unexpected that you could recover the original pairing at all. Second, the number that she gives to you looks nothing like the numbers on the original domino (no-one knows about your subtraction of the 6) yet you can identify her domino instantly.

Why does it work? By describing what is happening using a little algebra we can clarify it all in a minute, much more than working through any particular example where it continues to look like magic. Your friend has two numbers with both lying between 0 and 6 inclusive (and they may be the same). Call the number she selects to work with by the name a, and the other one b. Your instructions then tell her to compute: $(5a + 3) \times 2 + b = 10a + b + 6$. This final expression reveals what is going on. You take the 6 away, leaving you with $10a + b$; since a and b are single digits, this is equal to the two-digit number ab, enabling you to announce the result.

It is striking how the algebra strips away the inessential and allows you to see what is really going on. This contrasts with studying a particular case where the actual values of the numbers tends to distract and muddy the water.

Casting Out Nines

This is a diagnostic technique that you may have met in school that exposes errors in arithmetic. For addition the test is particularly simple. Suppose we have done the following sum and have come up with the answer on the right.

$$4398 + 1008 + 2129 = 7525.$$

To check the answer, sum all the digits on the left:

$$4 + 3 + 9 + 8 + 1 + 0 + 0 + 8 + 2 + 1 + 2 + 9 = 47;$$

if the answer is greater than 9, as in this case, keep on going until you have a single digit: $4 + 7 = 11; 1 + 1 = 2$. Our *magic number* is 2; if you have done the sum correctly then the number on the right will give the same magic number when this process of *casting out nines* is carried out there:

$$7 + 5 + 2 + 5 = 19; 1 + 9 = 10; 1 + 0 = 1;$$

but it doesn't, so we have made a mistake in the original sum. In fact the proper answer is 7535, which does give the correct magic number of 2.

Does this work every time? If the sum is correct, then the magic numbers will certainly match, and so if the magic numbers are not the same then the sum is wrong. However casting out nines can sometimes fail to diagnose an error. For example if you thought that

$$123 + 456 = 759;$$

casting out nines would not reveal your mistake as both sides in this case yield a magic number of 3. The correct answer is 579 and this type of example alerts us to a shortcoming of the test: since the outcome will be the same whatever order the digits are written in, casting out nines will never detect a transposition error such as this one where two digits have been recorded in the wrong order. This is the kind of error that you are likely to make when using a calculator or dialling a telephone number. When doing the sum yourself, a more common error is being 'one out' when summing the digits of a column, or failing to carry the correct digit to the next column. Casting out nines will pick up this error type every time—in the first example a carry of 2 from the unit's column was taken to the tens's column as a 1, leading to the magic number being out by 1, and so the error was detected.

Casting out nines is genuinely useful when doing multiplications. Reduce the *multiplicands* (numbers to be multiplied) to single digits, carry out the multiplication on them, and reduce again to obtain your magic number which you then check against that of your answer. For example, suppose you have come up with

$$462 \times 28 \times 49 = 638,864;$$
$$4 + 6 + 2 = 12, 1 + 2 = 3; 2 + 8 = 10, 1 + 0 = 1;$$
$$4 + 9 = 13, 1 + 3 = 4; 3 \times 1 \times 4 = 12; 1 + 2 = 3;$$

compare with

$$6 + 3 + 8 + 8 + 6 + 4 = 35, 3 + 5 = 8.$$

The magic numbers of 3 and 8 disagree so once again the given answer must be wrong. If we redo the sum we find the correct answer, 633,864 with the correct magic number of 3.

You should appreciate that you can always cast out nines as you go along—for example when casting out nines on the sum of the digits

$$7 + 7 + 9 + 6 + 5 = 34, 3 + 4 = 7;$$

we can mentally cast out as we proceed and get the same answer. The numbers in the thought process are then always kept below 20. In this case the mental sequence of numbers would be,

$$7 + 7 = 14, 1 + 4 = 5; 5 + 9 = 14, 1 + 4 = 5;$$
$$5 + 6 = 11, 1 + 1 = 2; 2 + 5 = 7.$$

In general the idea is to redo the original sum but with each number replaced by the single digit that results from the casting out process. Here is an example involving two operations:

$$113 \times (899 - 196) = 79,439;$$

the diagnostic sum we are led to if we cast out nines is $5 \times (8 - 7) = 5 \times 1 = 5$ and the number on the right (which is the correct answer) also has a magic number of 5.

Why does our test work? A clue is in the name—casting out nines. Let us look at one more example:

$$211 - 196 = 15$$

The sum is obviously correct: the answer has a magic number of 6 while casting out nines on the left leaves us with $4 - 7 = -3$: we are led to the conclusion that when casting out nines, 6 is the same as -3. This looks like an unwelcome development but it does provide us with a hint as to what is happening, for the numbers 6 and -3 differ by 9.

What casting out nines is really doing is checking whether both sides of your sum leave the same remainder when you divide by 9: if they do not they cannot possibly be equal. In these circumstances we say that both sides are equal modulo 9. Replacing one number in the calculation by a different one *that leaves the same remainder* does not alter that remainder. All this is true whether we are interested in remainders modulo 9, modulo 13, or modulo any number. The key thing about the number nine is that, when working in base ten, any number is equal, modulo 9, to the sum of its digits. This in turn is a consequence of the simple observation that one less than any power of 10 gives a number that is a string of 9's, and so is certainly a multiple of 9.*

This special property of the nine times table was the key to our Kangaroos in Denmark puzzle earlier, and is also the basis of the following trick where the magician claims that he can tell how many matches there are in a box just by listening to the rattle of the contents.[7] You (the magician) hand a box of matches to a member of the audience that contains a known number of matches—29 is a good number, as you will see. The audience participant is asked to take the matches out and replace as many matches as they wish, counting them as they go. You then ask them to add the digits of this number together and take that many matches from the box which they then return to you. You then shake the box and tell them exactly how many matches remain.

This is not so hard to do. Since any number is equal, modulo 9 to the sum of its digits, the number remaining will be a multiple of 9. For example, if the number of matches they initially place in the box is in the range from 20 up to 29, the box will have 18 matches when handed back to you. If they began with only a number between 10 to 19 inclusive however, there will be just nine matches in the box when you rattle it. It is not hard to tell whether there are 9 or 18 in the box just from the rattle so, with a little practice, you will be able to guess right every time. If you become more skilled, you might try a box with more matches. If you place up to 39 in the box, you will need to be able to tell the difference between a box with 27 matches and one with 18 or 9, but this is not too difficult.

Divisibility Tests

By a *Divisibility Test* for a certain positive whole number n we mean a way of deciding whether any given whole number m has n as a factor, or in other words, a way of telling whether m leaves a

---

7 My thanks to my colleague Dr Abdel Salhi for this one.

remainder of 0 or not when divided by *n*. If the answer is *yes* we say that *m* is *divisible by n*, or that *n is a factor* of *m*, or to put it a third way, that *m is a multiple of n*. For example, *m* = 36 is divisible by *n* = 6, while *m* = 56 is not, as the latter leaves a remainder of 2 upon division by 6. We can always answer a question of divisibility by carrying out the division in full so the test, if it is to be worthwhile, must generally involve considerably less work than doing the complete division sum.

1 and 10, 2 and 5

The base of our number system is 10. Making this choice was probably a mistake but it really is too late to turn back now. Which ever base you work in will provide very simple divisibility tests for the numbers which are factors of that base. The factors of 10 are, in mutual factor pairs, (1, 10), and (2, 5). If we worked in base 12 we would have a base with 1, 2, 3, 4, 6, and 12 as factors. The Ancient Bablyonians sometimes used base 60, a very round number that is even more rich in factors than 12. However ordinary arithmetic in this base would require learning multiplication tables up to 60×60, an idea that most of us would not be too keen on.

If we work in base *b*, and the number *n* is a factor of *b*, so that $b = kn$ say, then the last digit of the multiples of *n* follows a pattern, $n, 2n, 3n, \cdots (k - 1)n, 0$, as $kn = b$ will be written in base *b* as 10. The pattern of the last digit then repeats itself indefinitely as we pass on through all the multiples of *n* in base *b*. It follows that, working in base *b*, a number will be divisible by *n* if and only if its final digit is one of the digits $n, 2n, \cdots, 0$. That is to say, it is enough to check just the final digit for divisibility by *n*, and you can ignore the rest.

Applying this to our base 10 world we have that a number is divisible by 2 if and only if the final digit is one of 2, 4, 6, 8, or 0—that is to say a number is even if and only if the units digit is even. Similarly a number is divisible by 5 if and only if the final digit is either 5 or 0. The same idea applied to the pair of factors (1, 10) tells us that a number is divisible by 10 if and only it ends in a 0. I hesitate to mention divisibility by 1 as of course every number has 1 as a factor but, just to point out that the general argument works in this case also, we note that a number will be divisible by 1 if and only if the final digit is one of $1, 2, 3, \cdots, 9, 0$; of course every number passes that divisibility test!

The advantage of a duodecimal or base 12 system is apparent here. Working in that base, we could decide divisibility of a given number by any of the potential factors, 1, 2, 3, 4, 6 and 12 just by checking the final digit. For example, in base 12 the number 198 is $146_{12} = 1 \times 12^2 + 4 \times 12 + 6$, which is obviously divisible by 3 since this is true of its final digit. In base 10 this is not so obvious. (But see the divisibility test for 3 below.) However, whether or not a number is a multiple of 5 or 10 is less transparent when given in duodecimal representation: for example in base 12 we would write fifteen as 13_{12} ($= 1 \times 12 + 3$) and the factor of 5, although still present, is hidden from view.

4, 8 and 16

From here on the tests are a little less obvious. A number is divisible by 4 if and only if its final *two digits* represent a number divisible by 4. For example, 80,776,216 is a multiple of 4 by dint of the fact that 4 is a factor of 16, but 121,366 is not because 66 divided by 4 leaves a remainder of 2. The number represented by the final two digits is all that matters for if we take that away from the original number, we have a multiple of 100, which is certainly a multiple of 4. All that we need to decide then is whether those final two digits represent a further multiple of 4.

Note that this process does satisfy our criterion for a divisibility test for it reduces the problem from one involving a given number with an arbitrary number of digits, to a division involving a number with a fixed number of digits, that being, for this test, two digits.

To decide divisibility by 8, the test is much the same except that it is the *three* digit number at the end we must test. That is to say a number is divisible by 8 exactly when the number represented by the final three digits is a multiple of 8. For example you may care to verify that $a = 1, 894, 207, 376$ is divisible by 8, while $b = 3, 968, 844, 588$ is not. The rationale for this is along the lines of the test for 4: we only need check the behavior of the part of the number that comes from the last three digits as the rest, being a multiple of 1,000, is certainly a multiple of 8.

Note that, when it comes to 8, we cannot get by with just testing the final two digits. Indeed such a false test gives false results for both the numbers a and b above: 8 is a factor of a even though 8 is not a factor of 76, while 8 is not a factor of b yet 8 is a factor of the last two digit number, 88.

You will have noted a general similarity between the divisibility tests for 2, 4 and 8. For $2 = 2^1$ we check the final digit, for $4 = 2^2$ we check the last two digits and for $2^3 = 8$ it is the final three-digit number that is relevant. The pattern continues and can be justifed by extending the argument: a number is divisible by $2^4 = 16$ just when the same holds for the number formed by its last four digits. More generally a number will be divisible by a power 2^n of 2 exactly if this is true of the number you get by truncating the final n digits and working with that instead. The same observation holds good for powers of 5: a number is divisible by 5^n just when the number represented by the last n digits is divisible by the given power of 5. For instance multiples of $5^2 = 25$ are easy to spot as they are exactly the numbers ending in 25, 50, 75, or 00.

An example to test for a factor of 16 is $a = 5, 210, 224$. To be sure this is a particularly easy one as the last four digits are 0224; now $224/4 = 56$, and since 56 is also divisible by 4, we conclude that 224, and hence our original number a, is a multiple of $4 \times 4 = 16$.

3, 6, 9, 12 and 15

The divisibility test for 3 is quite a slick little trick. You may not guess but it is true that a number is divisible by 3 if and only if the sum of its digits is divisible by 3. For example, 792 is divisible by 3 as the sum of the digits is 18, while 721 is not a multiple of 3 as its digits add to 10.

This is a test that is truly easy to apply even for very large numbers as, although the sum s of the digits also may be a fairly big number, we can use the test on s itself. In other words, just as in the application of casting out nines, we can keep applying the procedure until we wind up with a single positive digit that represents the answer: if that digit is 3, 6 or 9 we have a multiple of 3, otherwise we have not. For example, let us test $a = 3, 406, 499, 617, 758$. The sum of the digits in this case is 69 and $6 + 9 = 15$, $1 + 5 = 6$, and so a is divisible by 3. As with the casting out nines technique, we can decide the question mentally by carrying out the process whenever the number in our heads exceeds 9; in that way the number we have in our minds is never more than 18. Performing this mental calculation on the number a above would see us doing the following mental process as we read the given number from left to right, probably using our finger to keep track of our place in the given number. In the explicit working below, the places where we pause to tidy up the number in hand and replace it by a single digit are written in brackets. Once that is done we continue reading in the digits of the given number from left to right:

$$3 + 4 = 7 ; 7 + 0 = 7; 7 + 6 = 13, (1 + 3 = 4); 4 + 4 = 8;$$
$$8 + 9 = 17, (1 + 7 = 8); 8 + 9 = 17, (1 + 7 = 8);$$
$$8 + 6 = 14, (1 + 4 = 5); 5 + 1 = 6; 6 + 7 = 13, (1 + 3 = 4);$$
$$4 + 7 = 11, (1 + 1 = 2); 2 + 5 = 7; 7 + 8 = 15, (1 + 5 = 6);$$

and hence a is a multiple of 3.

Since $6 = 2 \times 3$, a number is divisible by 6 if and only if it satisfies the divisibility tests for 2 and for 3 simultaneously. That is to say a number is a multiple of 6 exactly when its units digit is even *and* the sum of its digits is divisible by 3. For example, our number a above, being even is not only divisible by 3 but also has 6 as a factor. Similarly, since $12 = 4 \times 3$, a number is a multiple of 12 if and only if the number represented by its final two digits is divisible by 4 and the sum of its digits is a multiple of 3. I leave you to decide the question of divisibility by 12 for the numbers 477,168 and 861,774. Divisibility by 15 is also easily resolvable for a number will have $15 = 5 \times 3$ as a factor if and only if it ends in 5 or 0 and also passes the test for divisibility by 3.

These results, so easily obtained, show the power of the simple observation that many arithmetic operations can be broken into easy stages through use of factoring. In particular, if you don't like doing 'long' multiplications, they can often be avoided through multiplication by factors. If you know the multiplication tables of the factors of the multipliers by heart, you do not need to call on the long multiplication method: for example when you multiply a given number a by 84 say, long multiplication consists in arguing that

$$a \times 84 = a \times (80 + 4) = a \times 80 + a \times 4 = 10a \times 8 + a \times 4,$$

and so can be carried out as long as you know your 8 and 4 times tables.

An alternative is to do the sum as $a \times 12 \times 7$—provided you know your 12 (and 7) times table. If you don't trust your memory on the 12's you can do three little multiplications instead: $a \times 3 \times 4 \times 7$. In any case we see that long multiplication can be avoided up till the stage that the multiplier has a prime factor for which you do not know the multiplication table. For many people the first prime of this kind would be 13.

Last in our list of multiples of 3 there is 9 and, just as you might hope, a number is divisible by 9 if and only if the same is true of the sum of its digits. The justification for this is intimately related to that for the casting out nines diagnostic and will be explained shortly. Again you may like to convince yourself through examples such as $a = 59, 252, 085$ which is divisible by 9 (indeed it follows that a is divisible by $5 \times 9 = 45$) and 107,664 which, though a multiple of 3, fails the test for divisibility by 9. I leave it to the reader now to describe tests for divisibility by 18 and 36.

What makes the tests for 3 and for 9 work is the fact that any number is equal modulo 3 and modulo 9 to the sum of its digits.

In particular a number leaves a remainder of 0 when divided by 3 or 9 exactly when the same is true of the sum of its digits. This in turn is a consequence of the fact that any power of 10 leaves a remainder of 1 when divided by 3 or by 9 because a number that is a string of 9's is a multiple of them both.*

This is the basis of a devious little problem set for the teenage stars of the Mathematical Olympiad as a warm up exercise. You have a number a and you permute its digits in any way you wish to give another number b. Show that $d = a - b$ is never a prime.

This looks horrible: the difference d might, it seems, be anything and so how can we say much about its prime divisors? Certainly many of us would not know where to start and be left staring at the problem without hope of solution. A successful mathematician however has to maintain a playfull spirit in the face of a challenge and allow the question to lead where it will, even if the path looks unlikely to arrive at the destination sought. The one thing we can say about the numbers a and b is that the sum of their digits will be identical, and so a and b leave the same remainder when divided by 9. When we subtract one from the other, that remainder will vanish, leaving us with a number d that is a multiple of 9. And now we can see our way home: since d is a multiple of 9, it is certainly not a prime.

In the end we see that the bit about the prime was a red herring. If we had been told to explain why d had 9 as a factor, the problem would have been easy, even though that is a stronger conclusion than the one required. In a way the problem tests whether the candidate has the mathematical courage to put the peculiar conclusion aside for a moment and follow the mathematical signpost in the question. The moral is that students must trust their training and not be timid—easier said than done.

7, 11 and 13

There remain three awkward customers, 7, 11, and 13. These are primes that do not divide 10 and so their multiples are not too readily recognized when written in base 10. Eleven, being so close to ten, is the easiest to deal with. The divisibility rule for 11, although the most complicated so far, is easy enough to use.

If the sum of the digits of a number n, taken in order with alternating signs, is divisible by 11, then n is divisible by 11, and otherwise not.

For example, let us test $a = 56,518$ using our rule:

$$8 - 1 + 5 - 6 + 5 = 11;$$

which is a multiple of 11, and so 11 is a factor of our number a. Here we worked the digits in ascending order of their value—the opposite order yields the same outcome but with the opposite sign. The sign of a number does not affect its divisibility and so is unimportant in this context.

Another equivalent formulation of the test is as follows: let s be the sum of the digits in the even numbered places of a, and let t be the sum of the remaining digits. Then 11 is a factor of a if and only if 11 is a factor of $s - t$. The test number $s - t$ will either be the same, or the negative of, the test number given in the first version of the test according as the number a in question has even or odd length. In either case the same conclusion will be reached. Of course, in either version of the test, the test number may be negative. For example, if we take $a = 814, 396$ the test number in both versions is $(1 + 3 + 6) - (8 + 4 + 9) = 10 - 21 = -11$, again a multiple of 11. (You can always afford to ignore the minus sign.)*

As with the other sum of digits tests, we may re-apply the procedure for divisibility by 11 to the sum of digits that results until the number in hand is small enough to deal with by inspection. If we continue as long as possible one of two things will happen. Either we will finish with a non-zero single-digit integer, in which case the number is not divisible by 11 or, if it is a multiple of 11, we will end up with 0. For instance, if the alternating sum equals 154, applying the test to 154 would give $4 - 5 + 1 = 0$.

Here is an example that you can do quite easily that is well beyond direct division by a calculator: $a = 16, 193, 818, 284, 590, 452$;

$$s = (6 + 9 + 8 + 8 + 8 + 5 + 0 + 5) - (1 + 1 + 3 + 1 + 2 + 4 + 9 + 4 + 2) = 49 - 27 = 22; 2 - 2 = 0$$

and so a is divisible by 11.

A *palindrome* is a number that is the same when reversed such as 121, 181 and 2002. We can easily check that 181 is not a multiple of 11, but 121 and 2002 are. In fact every palindrome with an *even* number of digits has 11 as a factor because, as I am sure you can soon convince yourself, the sums s and t of the even and odd placed digits must be the same, so that their difference is 0, showing divisibility by 11.

Finally there is a digit based test that works for 7 and 13. In fact it also works for 11 but is more complicated than the one we already have for that number.

Let a be the given number. Start from the right and take each block of *three* digits and form the alternating sum, s, in the fashion for the test for divisibility by 11. The number a is divisible by 7 or 13 exactly when this is true for s. For example $a = 24,889,375$ is divisible by 7 but not 13. To see this we calculate the test sum s:

$$s = 375 - 889 + 024 = -490 = -70 \times 7;$$

but 490 is not divisible by 13, as is quickly checked.

Of course now that we have divisibility tests for 7 and 13, it is a simple matter to devise tests for the small multiples of these numbers: 14, 21, 28, ... and 26, 39, 52, ... respectively by coupling them with the tests for the other factors involved.

Let us close with a grand example. Is $a = 98,858,760$ divisible by 8008? Begin by factorizing the divisor: 8008 is a palindrome of even length so has 11 as a factor and obviously has 8 as a factor also: dividing gives $8008 = 11 \times 8 \times 91 = 11 \times 8 \times 7 \times 13$, and so we need to test a for divisibility by these four numbers. Since $760/2 = 380$, and 380 is divisible by 4, (because 80 is) it follows that a is a multiple of 8. We can test simultaneously for 7, 11, and 13 using the alternating sum:

$$s = 760 - 858 + 098 = 0,$$

and since 0 is certainly a multiple of all three numbers, we conclude that 8008 is indeed a factor of a.

Magical Arrays

A chapter on number magic is not complete without a word or two on magic squares and other magical arrays. The first magic square,

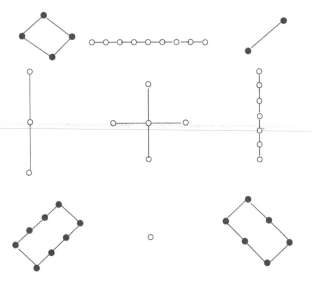

Figure 3.1. The first magic square.

the *lo-shu,* was presented to the Emperor Yu around 2200 BC by a divine tortoise along the banks of the Yellow River, or so the story goes. It was a square array of knots, black or white according as the number each represented was even or odd (see Fig. 3.1).

Each line of the magic square, whether it be horizontal, vertical, or diagonal, sums to the *magic constant*, which in this case is 15. In general a *normal magic square* of order *n* is a square array of the numbers from 1 up to n^2, with the property that all the lines sum to the same number, the square's magic constant.* It is easy to see that there are no 2×2 magic squares (apart from squares in

15	35	7
11	19	27
31	3	23

Figure 3.2. The lo-shu transformed.

which all numbers are identical). Indeed the lo shu is the unique 3×3 normal magic square, and so features each of the numbers 1 through to $3^2 = 9$.

Once we have an $n \times n$ normal magic square we can easily create infinitely many magic squares that look quite different from the original. We simply choose any two numbers we wish, a and b (they do not even have to be whole numbers) and replace each number k in the given square by $ak + b$. The effect of the multiplier a is to multiply each row total by a, while adding b to every number adds a total of nb overall to each row. If c was the old magic constant then the new square will have a magic constant of $ac + nb$. For instance if we choose $a = 4$ and $b = -1$ in the lo-shu we get the square of Fig. 3.2. In this case we have $c = 15$ and $n = 3$ so the new row sum is $4 \times 15 - 3 = 57$.

6	1	8
7	5	3
2	9	4

Figure 3.3. Complement of the lo-shu.

Another trick for generating a new normal magic square from an old one is to take its *complementary magic square,* the square that results when you subtract each number in the array from $n^2 + 1$. This will once again give you a square with every number from 1 to n^2 appearing exactly once. What is more, the line sums will be the same as before.* If we apply this idea to the lo-shu, we subtract each number from 10, and the square we recover is seen in Fig. 3.3. This is not a genuinely new magic square for it can be obtained from the original simply by rotating the square about its center through half of one full turn. The lo-shu is thus *self-complementary.* We can similarly find equivalent versions of any magic square by rotating it through any multiple of 90° about its center or reflecting it about one of its diagonals, or its vertical or horizontal axis. All the squares that result are considered to be copies

16	3	2	13
5	10	11	8
9	6	7	12
4	15	14	1

Figure 3.4. Dürer's magic square.

of the original as all eight versions could be seen by an observer as he moves around the original square or views it from the back.

The most famous example of a 4 × 4 magic square is that of the Albrecht Dürer (Figure 3.4), which appears in the top right hand corner of his engraving *Melancholia I*. Each line sums to the fourth magic constant of 34. Dürer's square has additional properties however, both mathematical and artistic. It features the additional symmetry that the sum of all the numbers in any of the four quadrants, as well as the sum of the four numbers in the center of the array, also sums to the magic constant, giving this magic square a special balance. Moreover Dürer cunningly adjusted the display so that the two central numbers in the bottom row give the year in which the picture was produced.

A simple connection between number squares and the magic constant arises as follows. Take all the numbers from 1 up to n^2 and display them in an $n \times n$ square as in the next picture. The following procedure always ends in the magic number for a magic square of the same size. Circle any number from the square and cross out all the other numbers in its row and column. Then circle a new number from the square and cross out any remaining in its row and column. Continue this process until you have chosen n numbers in all. The numbers you have selected will invariably sum to the magic number.* For example in Fig. 3.5 we have $n = 4$, and the diagram shows what happens when we follow the rules and choose, $11 + 14 + 5 + 4 = 34$, the magic number of Dürer's Square.

There is a simple trick, known as the *Siam Method*,[8] for producing normal magic squares of odd order. The idea comes about by first gluing the vertical edges of the square together to form a cylinder, and then the horizontal edges together to form a donut shape known as a *torus*.[9] We can describe this wrap around technique on a flat page however and we do it below for a 5×5 square. To the square adjoin fringe lines at the top and to the side and adjoin an additional shaded cell at the very top right hand corner

---

8 Named because it was brought to Europe around 1688 by De la Loubere, while serving as the envoy of Louis XIV to Siam.

9 Dürer's Square is also magical on a torus: when wrapped around any adjacent block of four adds up to the magic 34; eg. $3 + 2 + 15 + 14 = 34 = 5 + 9 + 12 + 8 = 16 + 13 + 4 + 1$.

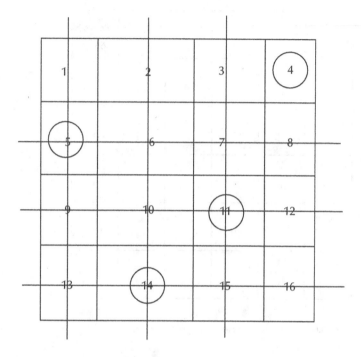

Figure 3.5. Selecting from each row and column.

as shown. Start by placing a 1 in the center of the top row, and write out the numbers from 1 up to 25 in this case, by proceeding upwards and to the right.

Exceptions occur however if this leads us out of the main square or into a cell that is spoken for through being already occupied by a number. In the latter case we simply drop down

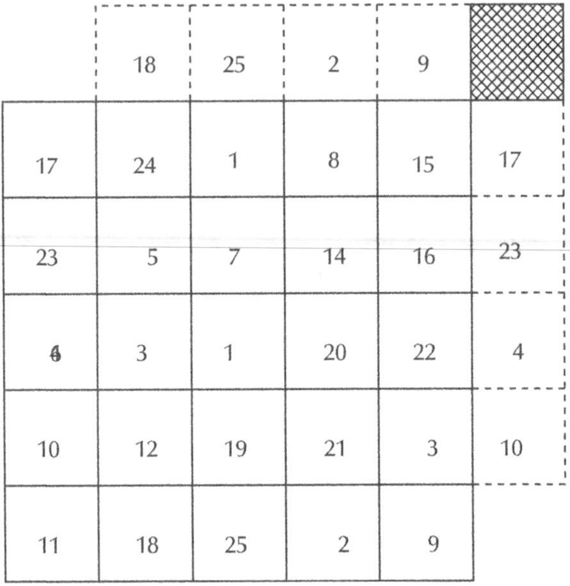

	18	25	2	9	
17	24	1	8	15	17
23	5	7	14	16	23
4	3	1	20	22	4
10	12	19	21	3	10
11	18	25	2	9	

Figure 3.6. Siamese magic square.

one square below the last one filled and continue as before. (The shaded cell is taken as occupied.) In the first case however we shift completely across the square, either from top to bottom or right to left as the case may be, and continue with the general rule. Applying this method produces the square of Fig. 3.6 with magic constant 65. The Siamese method applied to the 3×3 case yields the lo-shu, although reflected about its horizontal axis.

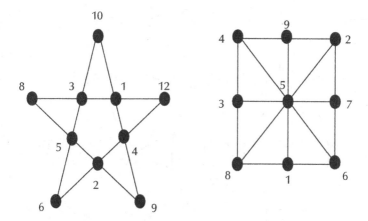

Figure 3.7. A magic pentagram and the lo-shu network.

Other Magic Number Arrays

We don't have to stick with squares.[10] Take any network of points and edges running between them and label the points with numbers. The network is then *magic* if every edge sums to the same number. For example the pentagram network in Fig. 3.7 has a magic constant of 24. Alongside it we again have the lo-shu, this time represented by a network of labeled points: each point

10 These are among many on the excellent resource http://mathworld.wolfram.com/
   These particular arrays feature in *Mathematical Recreations* (1979) by Joseph S.
   Madachy and appear here with permission of *Dover publications*, the book itself
   is no longer in print.

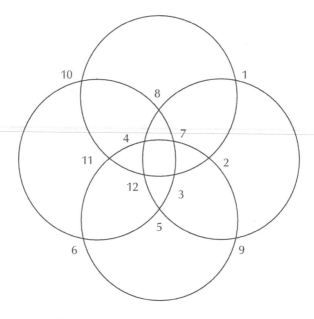

Figure 3.8. A set of four magic circles.

corresponds to a square and two points are joined by an edge if the corresponding squares are adjacent in a row, column, or diagonal.

There is no magic pentagram built on the numbers 1 through 10, from which it follows that there can be none featuring ten consecutive integers.

A set of *magic circles* on the other hand have numbers assigned to points of intersection and what makes them magic is a common sum around the circumference of each circle. In the example of Fig. 3.8 the magic constant is seen to be 39.

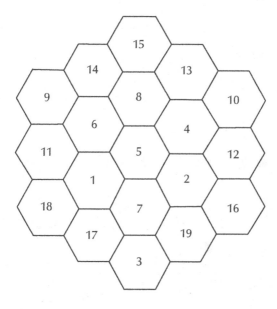

Figure 3.9. The magic hexagon.

Our final collector's item, shown in Fig. 3.9, is a magic hexagon: the numbers 1 through 19 are positioned so that every row in this bee hive sums to 38, irrespective of its length. Adams' hexagon, discovered in 1957, is unique: there is no other arrangement of consecutive counting numbers for any size hexagon that is magical. This may be why it took him most of his life to find it, having started his search in 1907.[11]

11 Apparently the magic hexagon has been discovered independently a number of times, including by Ernst von Haselberg in 1887.

chapter 4

Some Tricky
Numbers

Traditional number lore often focused on individual numbers thought to have special properties such as the perfect numbers mentioned in the first chapter. Another pair of numbers that captured the public imagination was 220 and 284, the first *amicable pair*, meaning that the sum of the factors of each summed to the other—a kind of perfection extended to a couple. Parted lovers would each carry an ornament decorated with one or other of these numbers as a token of their bond. Fermat (1601–1665) found others, such as 17,296 and 18,416 while Euler (1707–1783) found dozens of other amicable pairs. Surprisingly they all missed the small pair of 1184 and 1210, discovered by 16-year old Nicolo Pagnini in 1866. We can of course try to go beyond pairs and look for perfect triples, quadruples, and so on. These longer cycles are rare but do crop up.

We can start with any number, find the sum of its divisors less than itself, and repeat the process. The result is usually a little disappointing in that typically we get a chain that heads to 1 very rapidly, (reminiscent of the hailstone numbers to be seen later in this chapter), and with very few updrafts. For example, even beginning with a promising number such as 12, the chain is very short: $12 \rightarrow 16 \rightarrow 15 \rightarrow 9 \rightarrow 4 \rightarrow 3 \rightarrow 1$. The trouble is, once you hit a prime, you are finished. The perfect numbers are of course exceptions, each giving us a little loop, while an amicable pair leads to a two-cycle: $220 \rightarrow 284 \rightarrow 220 \rightarrow \cdots$. Numbers that lead to longer chains are called *sociable*. They were not studied at all until the 20th century as no-one had ever found any. Even today, no number that leads to a three-cycle has been found although there are now 120 known chains of length four. The first examples were found by P. Poulet in 1918. The first is a chain of length five:

$$12,496 \rightarrow 14,288 \rightarrow 15,472 \rightarrow 14,536 \rightarrow 14,264 \rightarrow 12,496$$

Poulet's second example is quite stunning and to this day no other cycle has been found that comes close to matching it: starting with 14, 316 we obtain a cycle of length 28. All other known cycles have length less than ten. To the present day there are no theorems on amicable and sociable numbers as beautiful as those of Euclid and Euler on perfect numbers* and the topic represents a nook in Number Theory that lies a little neglected. However, modern computing power has led to something of an experimental renaissance in this kind of topic.

On the other hand, numbers that arise of their own accord in enumeration problems are extensively investigated. There are many number types that are very special. Here I present a few along with some of the reasons that make them stand out. The binomial

coefficients, and the numbers of Catalan, Fibonnaci, Lucas, Stirling, and Bell, are significant because they count certain natural
collections. The primes however, are something else again and
deserve very special attention.

The most special class that arises in enumeration is that of the
binomial coefficients[12] or choice numbers as we have called them
previously. The binomial coefficient $C(n, r)$ is the number of different ways we may construct a set of size r from one of size n.
For example, as we saw in the first chapter $C(4, 2) = 6$, as there are
six pairs that can be chosen from a group of four. The binomial
coefficients can be calculated by means of the *Arithmetic Triangle*,
often also known as *Pascal's Triangle*, in honor of the 17th century
French mathematician Blaise Pascal (1623–1662).

Each number in the body of the triangle (see Fig. 4.1) is the
sum of the two above it. The triangle, which can be continued
indefinitely, gives the full list of choice numbers.

Number the lines of the triangle, beginning with 0 at the top.
Similarly number the positions within each row from left to right,
again starting with 0. To find the number of ways of selecting five
people from a group of seven, go down to the line numbered 7, and
then go to the number on that line numbered 5 (remembering to
start your count from 0): we see the answer is 21. You will note the
symmetry of each row: for example 21 is also the number of ways
of choosing two people from a group of seven. This is explained
by observing that when we choose the five from seven, we are
simultaneously choosing two from seven as well—the two being

---

12 So called because they are the coefficients that arise when the binomial expression
   $(1 + x)^n$ is multiplied out.

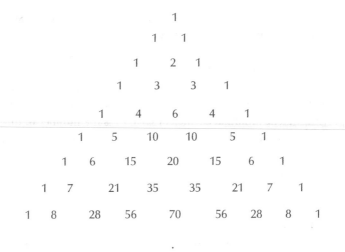

Figure 4.1. The Arithmetic Triangle.

the pair left behind. This symmetry argument of course applies to every row.[13]

The reason that the pattern gives the right answers is not hard to see. Each row builds from the one above it. We can see easily that the first three rows are correct: for example the 2 in the center of the third row tells us that there 2 ways of choosing a single person from

13 The Arithmetic Triangle seems to have been discovered and used in China around 1100 AD—it certainly forms the opening diagram of the classic mathematical work *The Precious Mirror* published in 1303 by Chu Shih-chieh.

a pair. The 1 that sits on top is saying that there is 1 way to choose a set of size 0 from the empty set. In fact there is 1 way of choosing a set of size 0 from any set which is why every row starts with a 1. Let us focus on the example just given—there are $21 = 15 + 6$ ways of selecting 5 from a group of 7 people. The 21 quintets split naturally into two types. First there are 15 ways to form a group of 4 from the first 6 people to which we may add the 7th person to form our fivesome. If we don't include the 7th person however, then we have to build a set of 5 from the first 6, and there are 6 ways of doing this. This illustrates how one row leads to the next: each entry is the sum of the two above it and this pattern propogates its way throughout the square.*

The triangle is rich in patterns. For example, if we sum the rows we get the doubling sequence 1, 2, 4, 8, 16, 32, $\cdots$: the sequence of powers of 2. In summing the row that begins 1, 8, 28, 56, $\cdots$ for instance we are summing the number of ways of choosing a set of size 0, 1, 2, 3 etc from a set of 7. In total this gives us the number of ways of choosing a set of any size from a group of 7, which is equal to 2^7 as, in general, a set of size n has 2^n subsets within it.* This is a point we touch on again in Chapter 7 when we consider infinite as well as finite sets.

Catalan Numbers

Every second row in the Arithmetic Triangle has a number sitting in the middle: 1, 2, 6, 20, 70, 252, 924, $\cdots$ These numbers are divisible by the consecutive counting numbers 1, 2, 3, 4, 5, 6, 7, $\cdots$ and the numbers that come about as we carry out these divisions, 1, 1, 2, 5, 14, 42, 132, $\cdots$ are known as the *Catalan numbers*. They

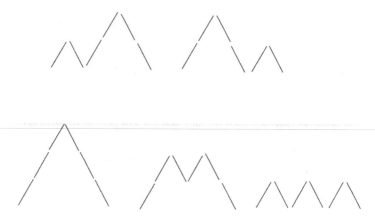

Figure 4.2. With 3 up and down strokes there are 5 mountain patterns.

arise in counting problems involving bracketing. The number of meaningful ways of arranging a collection of n pairs of parentheses is the nth Catalan number, or equivalently it is the number of ways we can draw n 'mountains' using n up strokes and n down strokes (see Fig. 4.2).[14]

The nth Catalan number also counts the number of ways that we can break up a regular polygon with $n + 2$ sides into triangles by means of diagonals that do not cross one another, and there are other interpretations along these lines. As with

14 For example (())() and ((())) are meaningful bracketings but ())(() is not: to be meaningful the number of left brackets must never fall behind the number of right brackets as we count from left to right. Equivalently, our mountains must never dive underground! In terms of the binomial coefficients, the nth Catalan number is $\frac{1}{n+1} C(2n, n)$.

binomial coefficients, there are formulas relating Catalan numbers to smaller Catalan numbers, which makes them amenable to manipulation.

Fibonacci Numbers

The Fibonacci sequence is one series of numbers that engenders wide fascination among the general public. The sequence runs as follows

$$1, 1, 2, 3, 5, 8, 13, 21, 34, 55, 89, 144, 233, 377, 610, \cdots$$

where each number after the pair of initial 1's is the sum of the two that come before. In this there is a similarity with the binomial coefficients in that each term is the sum of two previous ones in the sequence, but the method of formation of the Fibonacci numbers is simpler.

How does this sequence arise? It was first introduced in 1202 by Leonardo of Pisa, better known as Fibonacci, in the form of his now famous Rabbit Problem. A female rabbit is born and after two months reaches maturity and thereafter gives birth to a daughter each month. The number of female rabbits we have at the beginning of each month is then given by the Fibonacci numbers, for there is one rabbit at the beginning of the first month, and the second, but at the start of the third month she gives birth to a daughter so we then have 2 rabbits. Next month she has another and the month after that we have 5 bunnies as both mother and her eldest daughter are now old enough to breed. In general, at the beginning of each month thereafter, the number of *newborn* daughters equals the number of females we had *two* months ago, as only they are old enough to breed. It follows that the number of females we have at the start of each subsequent month equals the total of the previous month (Fibonacci's rabbits are immortal) plus the number we had the month before that. Therefore the rule of formation of the Fibonacci numbers exactly matches the breeding pattern of his rabbits.

Despite the fact that real rabbits do not breed in this contrived fashion, Fibonacci numbers arise in nature in a variety of ways including plant growth. The reasons for this are well understood but are related to more subtle attributes of the sequence.[15]

Of course it is possible to generate any number of Fibonacci-like sequences by slightly varying the rules of formation. The so called *Lucas Numbers* have precisely the same rule as the Fibonacci numbers only that the two initial values are taken to be 2, 1 in that order. The Lucas sequence bears a special relationship to the Fibonacci sequence and arises in various contexts in its own right. One aspect of the Fibonacci sequence noted by Lucas himself was the relationship the Fibonacci numbers bear to the Pascal triangle. As you can see below (Fig. 4.3), the Fibonacci numbers arise from summing the diagonals of the triangle in the manner indicated.*

Johannes Kepler (1571–1630) is most famous for discovering his Three Laws of Planetary Motion, which include the fact that planets orbit the sun in elliptical paths that sweep out equal areas in equal intervals of time. He was however a man who spent his life searching for patterns in nature, and he found one among the Fibonacci numbers, as we now explain.

15 Conway and Guy's, *The Book of Numbers* gives an explanation in terms of optimal properties of angles related to the golden ratio.

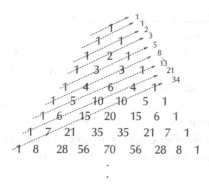

Figure 4.3. The Fibonaaci numbers and
Pascal's triangle.

The simplest number patterns are the *arithmetic* and *geometric progressions*. An example of the first is the sequence of odd numbers 1, 3, 5, ⋯ where the *difference* between any two consecutive terms is fixed, in this case the difference is 2. There are formulas for the nth term, and for the sum of the first n terms, which are easy to prove and to use. In this case the nth term is the nth odd number, $2n - 1$, and the sum of the first n odd numbers is n^2. In a geometric progression, we move from one term to the next, not by adding, but by *multiplying* by a fixed number. For example, the sequence of powers of 2 is an instance of a geometric progression, and identifying the nth term, and how to sum a geometric series are standard pieces of mathematics.

However the Fibonacci sequence is neither of these types. If we form the sequence of differences, because of the way the sequence is defined, we get $0, 1, 1, 2, 3, 5, 8, 13, \cdots$, that is we recover the Fibonacci sequence again except this time beginning at 0. This happens precisely because of the way the sequence is formed: the difference of two consecutive Fibonacci numbers is the one immediately preceding both in the sequence. Nor is the sequence a geometric progression as the ratio of consecutive Fibonacci numbers is not constant. What Kepler noticed however was that the ratio of successive terms does settle down to a limiting value. This near stable behavior of the ratio comes about quite quickly:

$$\frac{1}{1}, \frac{2}{1}, \frac{3}{2}, \frac{5}{3}, \frac{8}{5}, \frac{13}{8}, \frac{21}{13}, \frac{34}{21} = 1 \cdot 6190, \frac{55}{34} = 1 \cdot 6176,$$
$$\frac{89}{55} = 1 \cdot 6182, \frac{144}{89} = 1 \cdot 6180, \cdots$$

But what is the mysterious number, 1.618 . . ., which we see emerging? This number τ is known as the *Golden Ratio,* and it arises quite of its own accord in geometrical settings that look a world away from Fibonacci's Rabbits. For example, τ is the ratio of the the diagonal of a regular pentagon to its side, and serves to give this mysterious shape its peculiar symmetry and strength (see Fig. 4.4). Each diagonal meets another at a point that divides each into two parts that are themselves in the ratio $\tau : 1$. Pairs of intersecting sides and intersecting diagonals form the four sides of a rhombus (a 'square' parallelogram) $ABCD$ as shown. Where diagonals cross they form a smaller inverted pentagon of side length $\frac{1}{\tau^2}$ that of its parent.

Self-similarity is a characteristic feature of the Golden Ratio and is seen in the rectangle with sides of lengths τ and 1, for it displays the unique property that if we cut off the largest square

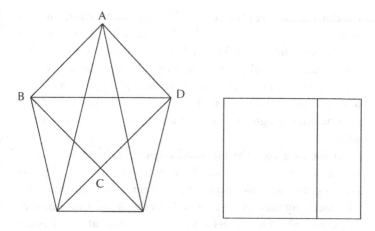

Figure 4.4. Pentagon and the Golden Rectangle.

we can (evidently a square of side length 1) the smaller rectangle that remains is a copy of the original. This figure is for that reason known the world over as the *Golden Rectangle.* A similar symmetry has long been exploited in the paper industry. The standard dimensions of an A4 sheet of paper are given as 297 by 210 mm. The corresponding ratio cancels to $\frac{99}{70}$, which is an approximation to $\sqrt{2}$ (to an accuracy of less than one quarter of a mm in terms of the sheet itself). This is sometimes called the *Lichtenberg ratio* as it was noted in 1786 by Georg Lichtenberg that, when folded down the middle on its longer side, a sheet whose sides were in the ratio of $\sqrt{2} : 1$ gives two smaller sheets exactly similar in proportion to the original. The upshot of this is that a larger sheet can be folded to produce similar smaller sheets, which is very convenient in the copying industry. For this reason, the standard A0 sheet, which is of dimensions 1189×841 mm has the required ratio and is almost exactly one square metre in area. Successive folding yields A1, A2, A3, A4 sheets etc, all of the same shape as each other and of the original. Or you may view things in reverse: placing two A4 sheets long-side to long-side gives a big A3 sheet with the same shape, only turned through 90° as the old long side is now the new short side.*

In the long run, the Fibonacci sequence behaves like a geo- metric progression based on the Golden Ratio. It is this prop- erty, together with its simple rule of formation, that causes the Fibonacci sequence to arise so persistently. World weary mathe- maticians are apt to express a degree of irritation at the excessive attention lavished on the number τ as some of its devotees imbue it with almost cosmic significance. It is genuinely special nonetheless and we shall see its influence more than once in the course of the rest of the book.

Stirling and Bell Numbers

Like the binomial coefficients, these come up a lot in counting problems and depend on two variables, n and r. The Stirling Number[16] $S(n, r)$ is the number of ways of partitioning a set of n members into r blocks (with no block empty, and the order of the blocks and within the blocks, is immaterial). For instance, the set $\{a, b, c\}$ can be partitioned into three blocks in just one

16 Strictly these are called *Stirling numbers of the second kind*. Those of the first kind, which are related, count something quite different, namely the number of ways we can permute n objects into r cycles.

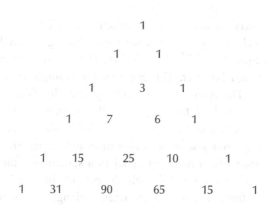

Figure 4.5. Stirling's triangle.

way: $\{\{a\}, \{b\}, \{c\}\}$, into two blocks in three ways $\{\{a, b\}, \{c\}\}$, $\{\{a\}, \{b, c\}\}$ and $\{\{a, c\}, \{b\}\}$, and into a single block in one way only: $\{\{a, b, c\}\}$; it follows that $S(3, 1) = 1$, $S(3, 2) = 3$ and $S(3, 3) = 1$. Since a set of n members can be partitioned in only one way into either 1 block or into n blocks, we always have $S(n, 1) = S(n, n) = 1$. If we draw up the triangle of Stirling numbers after the fashion of Pascal's Triangle we arrive at the array of Fig. 4.5.

Once again the numbers satisfy a *recurrence relation* in that each can be related to earlier ones in the array. Indeed, as with the binomial coefficients, each Stirling number can be got from the two above it, but it is not simply the sum. What is more, the

row symmetry we saw in the Arithmetic Triangle that generates
the binomial coefficients is not present in Stirling's Triangle. For
example, $S(5, 2) = 15$ but $S(5, 4) = 10$. The rule of recurrence is
simple enough however. The entry 90 for example is equal to
$15 + 3 \times 25$. This is representative of the general situation: to find a
number in the body of the triangle, take the two immediately above
it, and add the first to the second *multiplied by the number of the
position in the row you are at.* (This time, unlike the Arithmetic
Triangle, start your row count at 1.) In a similar way the entry
$S(5, 4) = 10 = 6 + 4 \times 1$. It is only the part of the rule in italics
that differs from that of the Arithmetic Triangle.* It is enough
however to make the study of Stirling Numbers considerably more
difficult to that of the binomial coefficients. For instance, we can
derive a simple explicit formula for each binomial coefficient in
terms of the factorials. Similarly, there is a formula for the nth
Fibonacci number in terms of powers of the Golden Ratio, but
nothing of the kind exists for Stirling Numbers which, it seems, can
only be computed recursively.* Each row of Stirling's triangle has
one hump, meaning that, reading from left to right, the numbers
increase to a maximum and then decrease down to 1—they never
go up then down, then up again. This may not surprise you in the
least but it is a fact that proves quite awkward to demonstrate in
general!

The sum of any row of the Arithmetic Triangle gives the corresponding power of 2— the number of subsets of a set of a given
size. Similarly, summing the nth row of Stirling's Triangle gives the
number of ways of breaking a set of n objects into blocks, and this
is called the nth *Bell Number*.

If, on the other hand, the n objects are identical, and so cannot be distinguished from one another, the number of ways of

splitting them up into blocks is a much smaller integer, known as the nth *partition number*. A partition corresponds to the number of ways of writing n as a sum of positive integers, without regard to order: for example we can represent 5 as $1 + 1 + 1 + 1+1$, $1 + 1 + 1 + 2$, $1 + 2 + 2$, $1 + 1 + 3$, $2 + 3$, $1 + 4$, or simply as 5. Therefore the 5th partition number is 7 (that compares to the 5th Bell number which, from the triangle above, is seen to be $1 + 15 + 25 + 10 + 1 = 52$).[17]

There is however a curious relation between *ordered* partitions of a certain kind and Fibonacci numbers. The number of ways of partitioning the integer n into an ordered sum of numbers all greater than 1 is f_{n-1}, the $(n - 1)$st Fibonacci number. For example, 8 has 13 such partitions:

$$2 + 2 + 2 + 2 = 2 + 2 + 4 = 2 + 4 + 2 = 4 + 2 + 2 = 2 + 3 + 3$$
$$= 3 + 2 + 3 = 3 + 3 + 2 = 2 + 6 = 6 + 2 = 3 + 5$$
$$= 5 + 3 = 4 + 4 = 8$$

and 13 is indeed f_7, the seventh Fibonacci number. This happens every time.*

Hailstone Numbers

Every now and then number patterns throw up an innocent looking problem that defies all analysis for a very long period. Many connected with primes will be met in the next section. In more

17 There is no simple exact formula for the nth partition number—there is a complex one and a beautiful limiting approximation due to Ramanujan: $\frac{1}{4n\sqrt{3}}e^{\pi\sqrt{2n/3}}$. A recursion for the partition numbers involving pentagonal numbers is due to Euler.

recent times, problems involving recursions such as those based on Fibonacci numbers have enjoyed a new lease of life as modern computing power allows us fast and direct investigation of their behavior for very long sequences indeed. The following example goes by several names, The *Collatz Algorithm*, the *Syracuse Problem*, or sometimes just the $3n + 1$-*problem* and it is simply the observation that, beginning with any number n, the following process always seems to end with the number 1. If n is even, divide it by 2, while if n is odd, replace it by $3n + 1$. For example, beginning with $n = 7$ we are lead by the rules through the following sequence:

$$7 \to 22 \to 11 \to 34 \to 17 \to 52 \to 26 \to 13 \to 40$$
$$\to 20 \to 10 \to 5 \to 16 \to 8 \to 4 \to 2 \to 1$$

And so the conjecture is true for $n = 7$ and indeed it has been verified for all n up beyond a million million. The sequences of numbers themselves that arise from these calculations behave like hailstones in that they rise and fall erratically over a long period but eventually it seems, always hit the ground. Of the first 1,000 integers more than 350 have a hailstone maximum height of 9,232 before collapsing to 1. All sorts of intriguing features can be discerned in graphs and plots based on the hailstone sequences reminiscent of other chaotic patterns that arise in math and physics. Is this a terribly important problem or not? The answer to that is not so clear but it certainly is hard—the late Paul Erdös (1913–1996), who perhaps knew more about numbers than anyone regarded it as a problem for which 'mathematics was not yet ready'. The Hailstone Problem may turn out to be a simple manifestation of a much more substantial problem, as we saw with Fermat's Last Theorem that was eventually solved as a consequence of another deep question in number theory, the so-called Shimura-Taniyama conjecture.

Be that as it may, there are some simple observations and reformulations. The conjecture is equivalent to saying that starting from any n, the rules will eventually lead you to some power of 2, because the powers of 2 are exactly the numbers that drop down to 1 without any updrafts. It would also be enough to prove that, beginning with any seed number n, the hailstone sequence will eventually hit a number $m < n$, for if that were *always* true, from m we would eventually reach a number smaller than m and, continuing in this way, the process would inevitably lead right down to 1. Typing 'Hailstone numbers' into your favourite search engine will provide you with a wealth of information, often intriguing, sometimes speculative, but generally inconclusive.

The Primes

The sequence of prime numbers is the most famous and important of all. As was proved in Euclid, the primes go on for ever. This really did require proof as we have no way of calculating arbitrarily large primes the way we can produce composite numbers or squares exceeding any given size. Despite millenia of effort, we are still searching for primes. The elusive nature of primes underpins public key encryption systems that feature later in the book.

To check that a number p is prime we have only to know that it is not divisible by any prime number up to $\sqrt{p}$. To see why this is so, suppose that $p = ab$, where a and b are proper factors of p, which is to say positive whole numbers greater than 1 but smaller

Figure 4.6. The Prime Sieve up to 30.

than p. Let us take it that the one we call a is no larger than b. It is not possible for *both* of a and b to exceed $\sqrt{p}$, for then ab would exceed p. The smaller factor a is certainly then no more than $\sqrt{p}$. Since a must itself have a prime factor, which is then also a factor of p, the first statement of this paragraph has been justified. We do have to check right up to the square root though—for example the only prime factor of 25 is 5, its square root.

The systematic calculation of all primes is, in principle, easy. The age old method is the *Sieve of Eratosthenes*.[18] We write all the numbers from 2 up to as far as you are prepared to go. Circle 2, and then delete all of its further multiples. Return to the start, find the first number that is not circled, circle it, and delete all of its multiples. Keep repeating this step until you circle a number that exceeds the square root of the largest number in your sieve. The numbers that you have not crossed out are then the required list of primes. In Fig. 4.6 for example we can stop sieving after we have crossed out the multiples of 5, as the next uncircled number is 7, whose square exceeds 30.

It is quite easy to see why this works. Clearly, while sieving, you will never cross out a prime. On the other hand every composite

---

18 Who is also famed for calculating the diameter of the earth in 230 BC through the difference in the length of shadows at Syrene and Alexandria at the Summer solstice.

number in the list will be crossed off as each number has been checked for prime factors as least as far as its square root, which is enough to decide the question one way or another.

A lot is known about the overall frequency of the primes. For any number n, there is at least one prime greater than n and no more than $2n$. For example, for $n = 5$ *Bertrand's Postulate* as this is known guarantees at least one prime in the range from 6 to 10 inclusive (and we see there is exactly one, namely 7, in this case). Although not a very sharp result (there are often many primes in the range specified by the postulate) theorems like this are by no means easy to prove. This kind of result is intrinsically interesting as it shows it cannot be quite true that the primes can be regarded as occuring randomly among the odd integers as, when we reach the number n, we are *absolutely sure* that we will meet at least one more prime before we get to $2n$. However, a lot of the modern theory of primes numbers takes the view that the global frequency of primes is, in a sense, random.

A conjecture similar to the Bertrand postulate that remains unresolved is whether there is always a prime between any two consecutive squares. A famous result of Dirichlet is that if we take two numbers with no common factor such as 3 and 8, then there are infinitely many primes in the arithmetic progression begin-nining at the first number and increasing in increments of the second. In this case that is the sequence 3, 11, 19, 27, 35, 43, $\cdots$ which therefore contains infinitely many primes. At first sight this may not look much tougher than the original result of Euclid that the sequence of odd counting numbers contains infi-nitely many primes, which is indeed the special case of the Dirichlet Theorem where the first number is 1 and the common difference is 2. However, all proofs of Dirichlet's result are deep and difficult. When it comes to primes you don't have to go far beyond an easy question before meeting an extraordinarily hard one.

On the other hand, no arithmetic progression can consist entirely of primes because the difference between two terms in an arithmetic progression is a fixed number while it is known that there are arbitrarily long gaps between prime numbers if we look far enough down the list of counting numbers.*

A simple observation though is that any prime, with the exceptions of 2 and 3, has the form $6n \pm 1$ because any number not of this form is either even or a multiple of 3. Indeed De Bouvelles in 1509 casually suggested that at least one of the numbers $6n \pm 1$ is always prime. His conjecture stands up to little scrutiny however: it first fails when $n = 20$, for $119 = 7 \times 17$ and $121 = 11^2$. It is quite simple to show nonetheless that there are infinitely many primes of the form $6n - 1$.*

The prime sequence displays peculiarities near the beginning that are never repeated. The number 2 is the only even prime, and the triplet of 3, 5, 7 is the one and only case of three consecutive odd numbers that are all prime, as for any number n, one (and only one) of $n - 2$, n, $n + 2$ is a multiple of 3.[19] Examining the prime sieve, we do however see pairs of *twin primes*, primes that differ by only 2: (11, 13), (17, 19), and it is not hard to find more. The *Twin Prime Conjecture* is that these pairs are never exhausted and go on appearing forever in the list. This is a major open problem that has remained defiant for centuries.

As has Goldbach's Conjecture, to be found implicitly in a letter to Euler dated June 7,1742: every even number greater than 2

---

19 Divide n by 3: if n is a multiple of 3 its neighbours are not, if the remainder is 1 then $n + 2$ has 3 as a factor but $n - 2$ does not, while if the remainder is 2 then $n - 2$ is the only multiple of 3 in our trio.

is the *sum* of two primes. For example $28 = 17 + 11$. A million dollar prize was on offer to anyone who could settle this conjecture in a two-year window that ended in March 2002. 'I'd want more than a million for that one', was one remark offered by a frustrated mathematician. Number theorists can become very irritated by Goldbach. There are some weaker versions that have been proved, and the conjecture has no doubt been verifed up to some extraordinarily large number.[20] It was shown by Schnirelman in 1939 that every even number from 4 onwards is the sum of no more that 300,000 primes. It was proved by Vinogradov around the same time that every sufficiently large odd number is the sum of three primes. (By *sufficiently large* we mean that this statement is true after some point, that point itself perhaps being unknown, but in this instance, particular bounds are known, although they are incredibly large.) A stronger result along these lines that seems to be true is that every odd number n is the sum of three primes with two of them the same—that is n is the sum of a prime plus twice a prime. Again though, this is unproved. A clever result that was proved by J.R. Chen is that every even number large enough can be written as a sum $p + m$, where p is prime and m has no more than two proper factors: that is m is prime or the product pq of two primes (that may be equal). This sounds tantalisingly close to the full Goldbach, yet it is not enough.

We still seem nowhere near deciding the original question. Hardy once testily observed that it is comparatively easy to make clever guesses; indeed there are many theorems, like 'Goldbach's

---

20 The current mark seems to be up to 6×10^{16} by Oliveira e Silva in October 2003: see the web page www.mathworld.wolfram.com.

Theorem', that have never been proved and which any fool could have come up with.[21]

It does not do to be too dismissive. The greatest mathematician of the 19th century, Karl Freidrich Gauss (1777–1855), once downplayed the importance of Fermat's Last Theorem: no nth power is the sum of two nth powers for any n greater than 2, by saying that he could list any number of simple questions about numbers that lay unresolved, so why focus on this one? However, the Fermat Theorem was well worth pursuing. It was solved in 1995 by Wiles in the affirmative but the question may yet have more to yield. However, the Goldbach Conjecture is still often dismissed as not belonging to the same category because, as the quip goes, prime numbers were never meant to be added up!

The overall frequency of the primes is quite well understood and is summed up by the observation of Gauss that $\pi(n)$, the number of primes up to the number n is, in the limit, equal to n divided by its natural logarithm (logarithms calculated to a special base number denoted by e). Gauss's guess, which actually claims something more complicated and precise, was proved in 1896 by Hadamard and independently by De La Vallee Poussin in the same year.

Since they represent so natural a sequence, it is almost irresistible to search for patterns among the primes. There are however no genuinely useful formulas for prime numbers. That is to say there is no rule that allows you to generate all prime numbers or even to calculate a sequence that consists entirely of different primes. There are some neat formulas but they are of little

21 As an example, we might suggest a kind of dual of Goldbach's Conjecture: every even number is the *difference* of two primes. This in turn is a special case of de Polignac's conjecture (1849) that infinitely many *consecutive* primes differ by any even number $2n$: when $n = 1$ this gives the Twin Prime conjecture.

practical worth, some of them even require knowledge of the prime sequence to calculate their value so that they are essentially a cheat. Some polynomial expressions such as $n^2 + n + 41$ are particularly rich in primes, yielding primes for many values of n. At the same time however, it is clear this one must fail when we put $n = 41$, as the result will have 41 as a factor. In general, it is not hard to show that no polynomial of this kind can yield a formula for primes.*

We can try recursions: start from 2 and keep doubling and adding 1. We do get a few primes: 2, 5, 11, 23, 47, but the next in the sequence is 95, and this method cannot be trusted either.

It is possible to devise tests for primality of a number that can be stated in a few words. However, to be of use they would need to be quicker, at least in some cases, than the direct verification procedure described above. A famous result known as Wilson's Theorem[22] tells us that a number p is prime if and only if p is a factor of $1 + (p - 1)!$ Despite being a concise statement, it is of no real use in identifying prime numbers. For example, to check that 13 is prime by Wilson would require us to verify that 13 is a factor of $1 + 12! = 479,001,601.$[23] Compare this to the labor involved in simply checking that 13 is divisible by neither 2 nor 3. Although Wilson's Theorem is not useful in prime verification, it has more than ornamental value and can be used, for instance, in the problem of determining which numbers are the sum of two squares. The theorem itself can be proved quite easily using the fundamental result known as Fermat's Lemma.*

---

22 Something of a misnomer, the result was first proved by Lagrange around 1770 and Leibnitz guessed it before anyone back in 1682.

23 But by applying the divisibility test for 13 on p. 47, we calculate $601 - 1 + 479 = 1079 = 13 \times 83$, and so we see that Wilson was right!

Lucky Numbers

Prime numbers are very special indeed but some of their behavior is mirrored in other trains of numbers. The sequence of so-called *lucky numbers* is generated by a kind of false sieve of Eratosthenes. We begin with the sequence of odd numbers only, $1, 3, 5, 7, \cdots$. The first lucky number is 3 (like the primes, we don't include 1 as part of the list, nor do we call it lucky). Put a circle around 3 and go through the sequence crossing out every third number, leaving us with a reduced list $1, 3, 7, 9, 13, 15, 19 \cdots$ The next uncircled number is 7, we circle it and then go through striking out every seventh number in the remaining list, and so on. As with the Prime Sieve, the numbers that are never be struck off form the *Lucky Sequence* that begins

$$3, 7, 9, 13, 15, 21, 25, 31, 33, 37, 43, \cdots.$$

Although lucky numbers lack the significance of primes, they share a similar frequency structure: the long term density follows the same limiting pattern, we can set up conjectures on twin lucky numbers and the counterpart of the Goldbach Conjecture and they seem to be true and, similarly to the prime case, they remain impervious to our current methods of proof. This suggests that many of the theorems and conjectures on prime numbers are in reality results about the sieving process and are not the exclusive preserve of the primes, although there does not appear to be any clear cut conjecture or formulation of this idea in the mathematical literature.

chapter 5

Some Useful Numbers

Percentages, Ratios, and Odds

One number type to which the modern world seems addicted is that of the percentage. There are good reasons for this however. We constantly need to refer to things such as growth, or to a certain proportion of the population, but at the same time we do not much like fractions—even talking in terms of decimal fractions is awkward in ordinary conversation. The notion of percentage is a simple idea that comes to the rescue. All of us prefer whole numbers to fractions and small counting numbers to large ones. We therefore choose to regard any measurable thing as consisting of 100 parts—as always we want to base our system on a power of 10, and 10 parts is a bit too coarse a measure to be really useful, and so 100 is used. One percent therefore is just 1/100 part of the object under discussion. To turn a fraction into a percentage, we simply need to multiply by 100.

Quartiles and *percentiles* we also hear much about. The first *quartile* of a data set is the point where $\frac{1}{4}$ of the data, that is to say 25%, lies below, while the rest sits above. In general, the kth percentile is the value where k percent of the data lies below, and the rest above.

Ratios are a way of comparing two fractions just through their numerators. For example, if the ratio of blue-eyed to brown-eyed people in the population is 4 : 3, we mean that for every three people with brown eyes there are four with blue. The relevant denominator is found by adding the numbers in the ratio so, in this case, the fraction of blue-eyed individuals would be $\frac{4}{7}$—at least if there were no other color eyes around. However the ratio of blue to brown to all other colors may be, let us say, 4 : 3 : 1 in which case the proportion of the blue-eyed would be $\frac{4}{8} = \frac{1}{2}$.

Any calculation involving ratios can therefore be turned into one of fractions and the ratio notation dispensed with. One place where ratios reign supreme however is in betting, especially on horse races. *Odds* of 2/1 mean the ratio of winnings to stake is 2 : 1—that is to say the punter will win two units for each one that he bets, provided he backs the winner. If these odds were fair (of course in general they are not, for otherwise bookmakers would not make money) this would mean the probability that his horse will win is $\frac{1}{2+1} = \frac{1}{3}$. That way, two thirds of the time he loses his one unit stake, but there is a chance of one in three that he wins 2 units (and get his stake back as well). His average losses will then be $1 \times \frac{2}{3} - 2 \times \frac{1}{3} = 0$, and overall he should tend to break even.

Odds of '2/1 on' means that the odds ratio is 1 : 2. In this case our punter is backing a strong favourite and the chances of his horse winning are judged to be 2 in 3, and so he only stands to gain one unit for every two that he puts at risk. If the book is fair, then the sum of all the numbers $\frac{r}{r+s}$, where r/s is a typical quote for a runner, summed over the entire field, will add up to 1. In practice of course, the sum will be somewhat deficient, and the amount that it falls short of 1 measures the bookmaker's overall advantage over his punters.

Some well known problems concerning simple ratios are intriguing. A classic arabic question involves a hunter who is running short of food and chances to meet upon two shepherds, one of whom has three loaves and the other five. They agree to share a meal, dividing the loaves equally among them. The hunter thanks the shepherds and goes on his way after paying them eight piasters for their bread. How should the shepherds divide the money?

The solution: each man ate $\frac{8}{3} = 2\frac{2}{3}$ loaves so that shepherd with five loaves gave $5 - 2\frac{2}{3} = \frac{15-8}{3} = \frac{7}{3}$ loaves to the hunter while the other shepherd only gave $\frac{1}{3}$ of a loaf. The ratio of one to the other is then $\frac{7}{3} : \frac{1}{3} = 7 : 1$. Therefore the first shepherd should get 7 of the 8 piasters with the other receiving the remaining 1 piaster. (The hunter paid $8 \div \frac{8}{3} = 3$ piasters/loaf but we were not asked about that.)

That one was pretty easy. How do we get started on the next?

A car is twice as old as its engine was when the car was as old as the engine is now. What is the ratio of the car's age to that of its engine?

The difficulty is that the one sentence containing all the information has not one, but two internal references, making it hard to untangle. It is deliciously confusing, but here is what is going on. There are three numbers involved—we begin by giving names to each. Let c be the age of the car, e the age of the engine, and d

the number of years ago when the car was as old as the engine is now. Another way of saying this, which is perhaps clearer, is that d is the difference between c and e, the current ages of car and engine (as $c = d + e$). Next write down what we know about how these numbers are related: as we have just said, $d = c - e$, for example if the car was 10 years old and the engine 7 then $10 - 7 = 3$ years ago the car was the present age of the engine. The other piece of information is reflected in the equation $c = 2(e - d)$, as the car is twice as old as the engine was, d years ago. Replacing d by $c - e$ in this little equation we obtain:

$$c = 2(e - d) = 2(e - (c - e)) = 2(e - c + e)$$
$$= 2(2e - c) = 4e - 2c$$

Adding $2c$ to both sides of this equations tells us that $3c = 4e$, so that the ratio of c to e is $4 : 3$.

Scientific Notation

This representation of numbers is so called because of its importance in scientific experiment as it constitutes the standard way of recording a measurement, especially of a quantity that is either very large or small, and only known to a certain degree of accuracy. For that reason it is often called *standard form* and consists of writing the quantity as a number between 1 and 10 multiplied by a suitable power of 10. For example, the speed of light in a vacuum is approximately $300, 000, 000$ m/sec, which in standard form is 3×10^8 m/sec. Standard form bypasses the inconvenience of having to write out long strings of zeros. For example, in chemistry there arises Avogardo's constant, the number of atoms in a mole. Its value to three significant figures is $6 \cdot 01 \times 10^{23}$.

Scientific notation is equally important when dealing with the very small. As an extreme example there is *Planck's constant* in quantum mechanics, the value of which, to five significant figures, is $6 \cdot 6262 \times 10^{-34}$ joule seconds. Written out in decimal notation this would be $0 \cdot 0000 \cdots 00066262$ where there are 34 zeros appearing before the significant digits, which themselves are known through experimental measurement.

Any calculation can be carried out in scientific notation and if the numbers involved feature large powers of 10, whether they be positive or negative, this is often the best way of expressing the computation in any case. For example

$$(3.14 \times 10^7) \times (6.21 \times 10^6) = (3.14 \times 6.21) \times 10^7 \times 10^6$$
$$= 19.5 \times 10^{7+6} = 19.5 \times 10^{13}$$
$$= 1.95 \times 10^{14}.$$
$$(2.4 \times 10^{18}) \div (1.1 \times 10^{11}) = (2.4 \div 1.1) \times 10^{18-11} = 2.2 \times 10^7.$$

The powers of 10, the indices as they are called, simply add in the case of multiplication, and subtract for division, as they keep track of the number of zeros in the sum. In a scientific calculation, the numbers involved are normally measurements so that the number of significant figures in the final answer cannot exceed that of any of the input numbers, for this would be claiming an accuracy in measurement that was not there. For instance, the exact value of $3 \cdot 14 \times 6 \cdot 21 = 19 \cdot 4994$, but this needs to be rounded to three significant figures in the answer given above.

The general method underlines the fact that we can carry out any multiplication and division provided we can do the same for numbers in the range of 1–10. This is the basis of calculation through logarithms. A once-and-for-all table was drawn up expressing each number between 1 and 10 as a power of 10—this was called the *common logarithm* of the number. The log of the answer to a multiplication could then be found simply by adding the indices throughout, and then inspecting the inverse table of anti-logarithms for the conversion of the answer back into standard form. The first table of logarithms to the base ten were compiled by Henry Briggs of Oxford in 1617 in conjunction with the founder of the idea, the Scot John Napier.

Meaning of Means

An average, or a *mean*, as it is also called is a way of summarising a collection of data by a single number. This number is meant to represent a typical value, and by doing so it gives an indication of where you might reasonably say the central point of the data lies. We look at the common means that arise in statisics, and then some of a more purely mathematical character.

Statistical Means

The *common mean* or *arithmetic mean* is just the ordinary average of the data set given. To find it, we sum all the numbers involved and divide by the total number of numbers we have.[24] For example, the mean of a die roll is $(1 + 2 + \cdots + 6)/6 = 21/6 = 3 \cdot 5$. Straight

---

24 Like the word media, data is plural—the singular in each case being medium and datum. For that reason we should say *these data* rather than *this data*. It is however a bit pedantic to fuss as the word datum is rather ugly, and not much used.

away we see one of the disadvantages of this average: if the data set consists entirely of integers the common average will generally not itself be integral. The mean may then be a number that is not in the original data set and could not possibly be. For example, suppose that the mean number of people in the average American household is $1 \cdot 9$. The meaning is that if we were to multiply the number of households by $1 \cdot 9$, we would arrive at the total number of people in America who live in a recognized household. It does therefore make some sense to talk about the mean size of a household, even if no household could actually be of that size.

Another misuse of the idea of the mean comes about when an experimenter treats two distinct populations as if they were one, and then calculates statistics based on this joint population. For example, suppose a very naive biologist failed to recognize the fundamental difference between spiders and insects and put them all under his single classification of *bugs*. If a sample of his bugs contained ten spiders and ten insects, he could count the total number of legs and divide by 20, to conclude that 'on average bugs have seven legs'. However, despite his average being a whole number, no bug has or could ever have seven legs, and our scientist has been led to his silly conclusion through confounding two quite different kinds of organism. Taking this nonsense to the extreme gives us jokes like 'the average human has one breast and one testicle'. Blunders of this kind are of course obvious, hence the joke, but the problem of not recognizing more than one distinct population within a group is common enough, and sometimes is not evident until after data has been gathered. When studying average heights of adults for instance, it is obviously important to distinguish between men and women, although it may be less important when other tastes and attitudes are quantified. When conducting sample surveys, money can also be wasted if the sampler is overly cautious and takes care to distinguish between two parts of the population that turn out to have no significant statistical difference with respect to the quantity she wants to estimate. The best way to stratify a sample often only comes with experience and cannot be decided before the fact.

One positive feature of the common average is that it takes account of the actual value of each datum and this is a real practical strength of this statistic. For example, in times past, a ship's captain relied heavily on measurements of star positions, compass bearings, and the like. A single measurement might be quite innaccurate so the practice developed of replacing it with the average of several measurements. It was intuitively felt that this average was unlikely to be far from the mark, whereas the value of a single measurement was very variable. This practical scientific principle has a sound mathematical foundation—the more measurements taken, the higher the probability that the average will lie very close to the true value, unless of course there is some systematic error in the process of measurement itself. If the captain's astrolabe was not calibrated properly, the average of his readings might settle down to a consistent value but it would still be wrong, simply because of the flaw in his instrument.

The equal respect that the mean shows for each datum can also be a weakness however, as the mean is sensitive to *outliers*, numbers that are far outside the range of the rest of the data. For example, if a household has seven people that includes just one smoker, who consumes a pack a day, then the mean number of cigarettes smoked by each member of the house per week is 20. This fact, although true, conceals rather than reveals the real picture, as nearly all the members of the household do not smoke at all.

More generally, the mean does not give a very good measure of the central point of a data set when the numbers involved are not symmetrically spread around the mean but, on the contrary, are rather skewed. One example is average earnings. Typically only about one-third of the population earn above the average, leaving about 67% of the people feeling hard done by. This comes about through people who earn above average salaries often earning way above the average. Without going into arguments on social justice, it is fair to say that the mean is not a good measure of central tendency when it lies too close to one end of the range. In this situation the *median* of the data set is perhaps a better indicator. The *median* is the value that half the data set lies below, and of course the other half above. It is the 50th percentile. In the case of earnings data, the median gives people a better idea of where they stand as compared to the 'typical' person in the population.

In order to calculate the median of a set of numbers, we need to arrange the data in order and then take the middle number. For example, the median of 3, 3, 5, 8, 12, 13, 13 is the middle number 8. If the number of pieces of data is odd, as in this case, the median will always be a number from the actual set of data. However, if we have an even number of pieces of information, we conventionally take the median to be mean of the middle pair. This is a bit of a fudge but normally makes little difference, at least for large data sets. It does however introduce the failing of the mean, which is that its value may not be a member of the data set itself and might even represent an impossible value in the real world.

For example, the median value of the numbers that can show on a roll of a die is $\frac{3+4}{2} = 3.5$, which is not integral, and is in fact the same number as the mean. In general, if the data is arranged symmetrically about the mean, the values of the mean and median tend to be quite close. It is only when the data set is biased towards one end that a significant difference between the two averages emerges.

The median is not a very nice statistic from a mathematical point of view. It is rather forgetful as it does not incorporate the full value of every number—for instance in our first example, if we were to replace the final value of 13 by 113, it takes no notice at all. The median can also be awkward to calculate as it takes a lot of work to rank a list of numbers in order: given one hundred numbers it is not too hard to calculate the mean but takes quite a bit more effort to find the median value.

A third common measure of central tendency is the *mode*, which is simply the most popular value. It has an advantage over our two previous means in that it is always a member of the data set, in fact it is the most common one. For instance the modal number of people per household will be, let us say 3—it will never involve a fractional number of people. The disadvantage of the mode is that it ignores all but the most popular value. It is appropriate however when the data set does not involve numbers. For instance, if a group of children consisted of 30 with blue eyes, 18 with brown, and 6 with some other color, it is of some use to say that the modal color is blue. It makes no sense however to try to calculate a mean or a median eye color for the group.

If the data is quite uniformly spread, the mode is of little or no interest. For instance, for the values on the face of an ordinary die, each number from 1 through 6 comes up once. All six possibilities are equal modal values.

Another example of a data set where emphasis on the mean and median is somewhat deceptive is noontime cloud cover, expressed as a percentage to the nearest whole number, in a certain location. The average portion of the sky filled with cloud may be 50%. However, at many places of the earth's surface one of the two extremes is the norm—the sky is either completely clear, or there is total cloud cover. Partial cloud cover does happen of course but is relatively unstable, and often represents just a brief transition from one extreme to the other. In this case the modal value will be either 0 or 100, but not some number in between. For this data type, none of the averages of mean, median, and mode give a very good indication of a typical value of the variable.

Mathematical Means

Having seen that there are at least three different kinds of averages, it is fair to examine the question of what constitutes an average, before we try to identify any more. An *average* of a set of numbers is itself a single number that is guaranteed to lie between the minimum and maximum of the set. This is the very least it must do. In addition, in order to be a good average, it should in some interpretation represent the center of the data set.

Since the arithmetic mean A of a and b (with $a \leq b$) sits halfway between them, the three numbers a, A, b form an arithmetic progression, so the gap between each pair of consecutive numbers in the sequence is the same. Indeed A is the one and only number between a and b that yields such a progression. In the light of this, we may define the *geometric mean* of a and b as the number G such that a, G, b are in *geometric* progression. For this to make sense there must be a multiplier r such that $ar = G$ and $Gr = b$. The appropriate value for G is $\sqrt{ab}$.[25] For example, if $a = 4$ and $b = 9$, we find that $A = 6 \cdot 5$, while $G = \sqrt{36} = 6$. In general, for n numbers $a_1, \cdots, a_n$, their geometric mean is the nth root of their product.

Another way of passing from the arithmetic to the geometric mean is arrived at by taking the definition of A and replacing each arithmetic operation by the next one up in the hierarchy—that is to say we replace addition by multiplication, and replace multiplication by $1/2$, by exponentiation to the power $1/2$, that is to say the square root. The expression for A then becomes that for G. The geometric mean of two numbers a and b has the additional interpretation that the square with side length G, has the same area as the rectangle of dimensions a and b.

A third curious mean that arose in the musical considerations of the Pythagoreans is the *harmonic mean* H of a set of numbers. Three numbers $a < H < b$ are said to be in *harmonic progression* if their reciprocals form an arithmetic progression, which is to say that $1/H$ lies midway between the reciprocals of a and b. We then say that H is the *harmonic mean* of a and b. A little algebra allows us to conclude that

$$H = \frac{2ab}{a + b} \text{ which comes from, } \frac{2}{H} = \frac{1}{a} + \frac{1}{b}.*$$

---

25 Substituting the first equation into the second, we find that $ar^2 = b$, so that $r = \sqrt{\frac{b}{a}}$, which gives $G = ar = a\sqrt{\frac{b}{a}} = \sqrt{ab}$.

For example, if once more $a = 4$ and $b = 9$, we obtain $H = \frac{72}{13} = 5\frac{7}{13}$. Again the notion of harmonic mean also extends to more than two numbers: the harmonic mean of n numbers is n times their product divided by their sum.

A basic fact concerning the three means is that if $a < b$ then their relative order is always the same:

$$a < H < G < A < b.$$

(Of course if $a = b$, any mean of a and b must equal their common value.) This in turn has an interpretation that allows us to recover the classical method of Heron of Alexandria (AD 150) that affords calculation of $\sqrt{2}$ and other square roots to any required degree of accuracy.*

The ancient Babylonians recognized these three means, and the Greeks added another seven means m to the list based upon rules concerning the ratios of the numbers a, b, and m. The name of harmonic attached to our third mean is due to the Greek Archytas, the ruler of Tarentum in the 5th century BC. He is noted for being an invincible general, of being kind to children for whom he invented all manner of toys, and his insistence on the rule of numbers in affairs of state. To him is attributed the *quadrivium,* the four branches of the mathematical curriculum—arithmetic (numbers at rest), geometry (magnitudes at rest), music (numbers in motion), and astronomy (magnitudes in motion). These together with the *trivium*: grammar, rhetoric, and dialectic, came to constitute the seven liberal arts. His interpretation of music is particularly astute, and came from his conviction that pitch was due to the varying rates of motion resulting from the flow causing the sound.

The idea of the reciprocal of the sum of reciprocals occurs quite commonly in other problems to do with rates. For example, let's work out this brain teaser, which is a typical example of this kind.

The cold water tap fills the bath in six minutes but the hot water tap takes eight. How long does it take to fill the bath using both taps at the same time?

Taking the volume of the bath to be one unit, the cold and hot taps respectively fill at rates of $\frac{1}{6}$ and $\frac{1}{8}$ units/minute. The combined rate then is the sum of these two unit fractions, $\frac{1}{6} + \frac{1}{8} = \frac{4+3}{24} = \frac{7}{24}$ units/minute. Using an index of -1 to denote reciprocation, that is to say, the turning of the fraction upside-down, the required time is therefore

$$\left(\frac{1}{6} + \frac{1}{8}\right)^{-1} = \frac{24}{7} = 3\frac{3}{7} \text{ minutes}$$

as if we fill the bath at $\frac{7}{24}$ units per minute for $\frac{24}{7}$ minutes, the total volume of water delivered will be $\frac{7}{24} \times \frac{24}{7} = 1$ bath tub full.

To the nearest second, the time taken to fill the bath is 3 minutes and 26 seconds.

Another example where a quantity adds through its reciprocals is that of resistance in a parallel circuit.

This is how a simple circuit works. In Fig. 5.1(a) we have a power source, let us say a battery, the ends of which are connected by a wire so that current flows around the circuit from $(+)$ to $(-)$. The flow of current is inhibited by a resistor placed in the path of the current as shown.

What is constant in this situation is the *voltage*, V, of the battery which measures the energy borne by each of the charged particles,

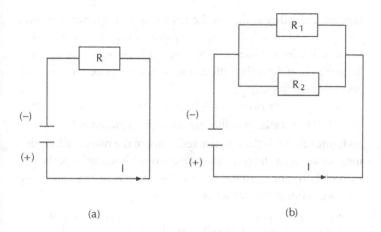

Figure 5.1. Resistance in a simple and in a
parallel circuit.

known as electrons, as they flow around the circuit. The voltage,
which for convenience we shall take as 1 volt, is equal to the product of the current, I and the resistance, R. In other words I and R
are *inversely proportional* to one another, as one increases the other
decreases in such a way that their product is the constant V; in
this case this gives $I R = 1$. (The standard units for each is *amps* for
current, and *ohms* for resistance.)

If we set up two resistors of respective values R_1 and R_2, one
after the other, we say that they have been connected *in series*. The
total resistance in the circuit is then $R_1 + R_2$ and the current in this
case would be the reciprocal of that sum.

The more interesting situation is when the resistors are
mounted in parallel (see Fig. 5.1(b)). The current will now happily

flow around either branch of the circuit so each behaves independently of the other. (Although the paths of the currents meet up again, the wire is taken to have negligible resistance of its own so there is no retarding effect due to the two rivers of current becoming confluent.)

The current carried by each branch is respectively $I_1 = \frac{1}{R_1}$ and $I_2 = \frac{1}{R_2}$. The total current flowing through the circuit is $I = I_1 + I_2$ and, since $R = \frac{1}{I}$ is the effective resistance of the two parallel resistors, we arrive at the conclusion referred to above, as by replacing each of I, I_1, and I_2 in the previous equation by the reciprocal of the corresponding resistance we get:

$$\frac{1}{R} = \frac{1}{R_1} + \frac{1}{R_2} \quad \text{or in other words} \quad R = \left(\frac{1}{R_1} + \frac{1}{R_2} \right)^{-1}.$$

In summary, in series resistances add, but when connected in parallel it is the reciprocals of the resistances that add.

chapter 6

On the Trail of
New Numbers

The strategy of this chapter is to take the *Natural Numbers* as they are called for granted, beginning with the number zero, 0, 1, 2, $\cdots$, and to follow where the ordinary questions and operations of arithmetic lead. This returns us to the vexing question, what is 3–4? One response might be that there is no answer to this question (recall the argument about the ducks) and we should be content to leave it at that. Any attempt to invent new numbers in order to satisfy our numerical appetite will be doomed to failure and confusion as the subject inherently lacks meaning.

This is a reasonable enough opinion, but it is only an opinion. Like any prediction in the face of uncertainty, its worth can only be judged through being put to the test. What is more, the argument stands vulnerable to refutation along lines of its own making. There are things to be dealt with in the world that have a numerical flavour going beyond what we need to count ducks. Take for instance debt—a human invention to be sure but one we all feel to be real and something that truly has to be reckoned with. The arithmetic of debt, which is after all negative money, requires that we know how to deal with both positive and negative numbers.

Putting concerns about debt to one side for the moment, it has long been established, for the best part of two centuries, that in order to allow mathematics to work for us in the wonderful ways that it can we should have no inhibition about the kinds of numbers that should be employed in calculations. Mathematics is trustworthy, and even if a calculation takes you to places in the number world that you had no intention of visiting, the numbers will not lead you astray, even if the meaning of the answers is not immediately apparent. This kind of talk requires examples to become convincing. It is not possible of course to clarify the role of really sophisticated mathematics in a few sentences but there are simple instances that serve to make the point quickly enough. Let us take a counting problem that involves only ordinary positive numbers. It is indeed a question about money.

How many ways are there of changing $1 into coins?

The coin types available are the usual 1c, 5c, 10c, and 25c coins. There are a large number of solutions including 4 quarters, 10 nickels and 5 dimes, and so on. The problem is of a type usually described as *combinatorial*, one that involves a finite count, and so, in principle could be solved by listing all the possibilities and counting them up. In practice however this would be very, very difficult, even for this 'toy' problem as mathematicians are apt to call simple examples. Many applications of modern applied discrete mathematics are of this combinatorial variety where we are called upon to cope with an enormous number of possibilities. In practice, a combination of clever mathematics and well designed software have to be brought to bear in order to tame them. For this reason combinatorics has been called the art of counting without actually counting.

The previous type of problem may be solved using techniques that go only a little beyond school mathematics. One algebraic method uses summation techniques of geometric and related series, followed by the method of *partial fractions,* which is a standard way of breaking fractions involving a symbol for the unknown x into a sum of separate but simpler fractions. The answer to the question can them be gleaned by plugging in the numbers that describe this particular problem. The key point however is that, in order to work through the algebra that arises, full use needs to be made of not only negative numbers and fractions but sometimes complex numbers as well—numbers that involve the 'imaginary' number i, the square root of minus one. The final answer, of course, is a simple positive integer, but the path that leads to its calculation involves numbers of very different kinds. If we refused to use complex numbers out of stubborness disguised as some kind of bogus philosophical objection, a solution to a whole range of important problems would remain forever out of reach.

The counting numbers are just the tip of the Number Iceberg. This tip is of course the first part we discover, and for a time we might believe there is no more to the iceberg than the tip, especially if we remain reluctant to look below the waterline. However one of the great achievements of the 19th century was the full realization that the true domain of Number is not one, but rather is two-dimensional. The plane of the complex numbers is the natural arena of discourse for much if not most of mathematics.

Pluses and Minuses

The *integers* is the name applied to the set of all whole numbers, positive, negative, and zero. This set is often symbolised by the letter $\mathbf{Z}$ and is therefore infinite in both directions:

$$\mathbf{Z} \{\cdots -4, -3, -2, -1, 0, 1, 2, 3, 4, \cdots\}.$$

The integers are often pictured as lying at equally spaced points along a horizontal line, naturally called the *Number Line*, in the order indicated. This mathematical representation is more than just a metaphor and has real uses. It is so pervasive that anyone with any familiarity with the integers has the picture of the number line pass before their mind's eye as soon the topic is mentioned. It is well to remind ourselves perhaps that it is a relatively modern way of thinking about numbers—it is not an image that would have occurred to Euclid or Archimedes.

The additional rules that we need to know in order to do arithmetic with the integers can be summarized as follows:

(a) to add or subtract a negative integer, $-m$, we move m spaces to the left in the case of addition, and m spaces to the right for subtraction;

(b) to multiply an integer by $-m$, we multiply the integer by m, and then change sign.

In other words, the direction of addition and subtraction of negative numbers is the opposite to that of positive numbers, while multiplying a number by -1 swaps its sign for the alternative. For example, $3 + (-5) = -2$, $3 \times (-5) = -15$, and $(-1) \times (-1) = 1$.

You should not be vexed by the last sum. It is for example reasonable that multiplying a negative number by a positive one yields a negative answer: when a debt (a negative amount) is subject to interest (a positive multiplier greater than 1) the outcome is greater debt, that is to say a larger negative number. That multiplication of a negative number by another negative number should have the opposite outcome, that is a positive result, then looks consistent. That double negatives yield a positive is common in other circumstances such as that of ordinary language. You could fall back on the objection that multiplying two negatives should not have any outcome at all as it does not lend to easy physical interpretation. However, for reasons explained above, you should not cling to such misgivings if you want your mathematics to progress. The fact that the product of two negative numbers is positive can readily be given formal proof based on the assumptions that we want our expanded number system of the integers to supersede the original one of the natural numbers, and that the augmented system should continue to obey all the normal rules of algebra.* Nor should you confuse yourself with meaningless interpretations. It is true that you cannot multiply one debt by another debt and get a credit, but that is because it makes no sense to multiply one pile of money by another pile of money in the first place, whether you regard the piles as credits or debits.

Fractions and Rationals

In a similar way that subtraction leads to the negative numbers, the operation of division also leads us out of the set of natural counting numbers into the larger class of fractions. However, the nature of the new arithmetic we encounter is of quite a different character. Although people often feel uncomfortable when dealing with the negative integers, the arithmetic rules of the full set of integers, as described above, are not at all complicated and hardly go beyond what we need to do arithmetic with purely positive numbers. The need to deal with fractions however seems to have been recognized from the beginning but the best way to set up the arithmetic is not so clear: the arithmetic of fractions is comparatively messy.

The Ancient Egyptians were only happy with fractions that were simple reciprocals of whole numbers, $\frac{1}{2}$, $\frac{1}{3}$, $\frac{1}{4}$ etc. A fraction such as $\frac{3}{4}$ was not thought of as a meaningful entity in its own right and they would record this quantity as the sum of two reciprocals: $\frac{3}{4} = \frac{1}{2} + \frac{1}{4}$. They did admit one minor exception however in introducing a special symbol for $\frac{2}{3}$. This insistence on only recognizing what we call unit fractions is novel and itself throws up interesting problems. It was however a wrong turn—they were rejecting fractions with numerators other than 1 for no sound reason. It is not even obvious that it is possible to write any fraction as the sum of a number of *different* unit fractions, which is what they insisted on. It can however always be done as they seemed to have appreciated. This would have had the unfortunate effect of confirming them in their view that this was the right way to go about things. The History of Numbers is littered with instances like this where a culture insists on handicapping itself in one way or another by turning its back on new types of numbers that were waiting in the wings ready to work for them. We should not be too judgemental though as our own adherence to base ten arithmetic is a further example of a prejudice of this type, although perhaps a less harmful one.

If you wish to find an Egyptian decomposition of a fraction such as $\frac{9}{20}$, you need only subtract the largest unit fraction you can from the given number, and repeat this process until the remainder is itself a unit fraction. This will always work, and the number of fractions involved never exceeds the numerator of your original fraction. This is because, at each stage, the numerator of the fraction that still remains is always less than the previous one: not obvious but true.* If you try this example you will get the decomposition:

$$\frac{9}{20} = \frac{1}{3} + \frac{1}{9} + \frac{1}{180}$$

This greedy approach of always subtracting the largest unit fraction available does work but may not yield the shortest decomposition there is, as we can see even in this case as $\frac{9}{20} = \frac{1}{4} + \frac{1}{5}$. After 5,000 years, the problem of how to find the shortest Egyptian decomposition of a fraction remains open.[26]

Multiplication of fractions is easy: we just multiply the numerators (top lines) and denominators (bottom lines) together. This is easy to see for unit fractions as, for example, $\frac{1}{3} \times \frac{1}{4} = \frac{1}{12}$ because we interpret the multiplication to mean one third *of* one quarter. If the numerators are not just 1 we can still see what is going on:

$$\frac{2}{3} \times \frac{3}{4} = 2 \times 3 \times \frac{1}{3} \times \frac{1}{4} = 6 \times \frac{1}{12} = \frac{6}{12} = \frac{1}{2}.$$

(It would have been quicker, as is always the case, to cancel the product down before doing the multiplication, but the example as given serves the purpose of revealing why fractions multiply as they do.)

A division undoes the effect of the corresponding multiplication. Dividing by 2 has the opposite effect to multiplying by 2 so we do the division by multiplying by $\frac{1}{2}$. In general, we carry this

26 This two-fraction decomposition of $\frac{9}{20}$ can be found however through use of the technique of the Akhmim papyrus.*

over to other divisions so, for example, to divide by $\frac{3}{4}$ we multiply by its reciprocal $\frac{4}{3}$. In words, to divide one number by a fraction, we multiply by the fraction inverted. It is addition (and subtraction) of fractions that leads to some difficulty. Two fractions with the same denominator can be added and subtracted just by adding and subtracting the numerators but when the denominators are different we are left facing a genuine incompatibility. It is akin to adding two numbers in different bases: this cannot be done until after the numbers have been converted to a common base. The same with the fractions: we have to convert each of them to a common denominator and then perform the addition or subtraction as the case might be. We usually keep the numbers as small as possible by finding the *lowest common denominator*. (Curiously this phrase has entered ordinary language and generally means the worst aspects of our makeup that we share in common: in arithmetic however, low numbers are good and high ones are more difficult to handle.) This is the least common multiple of the two denominators c and d. We can though, if we choose, find a common denominator just by multiplying c and d together. The smallest possible common denominator of c and d is cd/h, where h is the highest common factor of c and d.

The method just described has filled many a school morning and certainly works. It is still however unsatisfactory and that accounts for the fact that it is not much called upon because not only is the method awkward, but the answer it provides often is not that useful in practice. People tend to hanker for decimal representations. This is not just prejudice. Much of the importance of the final answer lies in the user being able to say if the answer is greater or smaller than a given quantity, (Do we have enough money to buy all this?) and indeed sometimes the user simply wants to have a clear feeling for the size of the quantity involved. Vulgar fractions don't always provide that assurance. For example, let us add the three unit fractions with denominators 3, 4, and 5.

$$\frac{1}{3} + \frac{1}{4} + \frac{1}{5} = \frac{(4 \times 5) + (3 \times 5) + (3 \times 4)}{3 \times 4 \times 5} = \frac{20 + 15 + 12}{60} = \frac{47}{60}.$$

We have the answer but do we feel much the wiser? The Ancient Egyptians would, after all, have left the sum as it was in the first place. Are we any better off having done the addition?

Anyone confronted with this final fraction will probably want to change it into its decimal equivalent: 0.783333 …. Why do we feel the need to do this? We already have the exact answer so what advantage is there in converting to a decimal, expecially a non-terminating one like this that goes on forever?

Psychologically we only have a feel for a limited collection of numbers. An answer in fractional form is only of immediate use if we can compare it with a number we feel we know well. In this instance we can see that the answer is bigger than one half, which is telling us something, but to find how much bigger we would have to do a subtraction sum involving much the same amount of work as the one just performed. If however we have the answer in decimal form we can see immediately it is not only more than one half, but more than three quarters (= 0.75) and we can even write down exactly how much bigger it is than this fraction: 0.0333 … is the excess over three quarters. In treating fractions in decimal form we are carrying over our base ten system for whole numbers to the realm of fractions, along with all its advantages of uniform presentation and ease of comparison. For example, when working with the decimal representations it is clear at first sight that $\frac{19}{24} = 0.791666 \ldots$ is greater than $\frac{47}{60} = 0.78333 \ldots$ but that is not clear when presented with both these numbers as vulgar fractions.[27]

The use of decimal fractions is found in ancient China and medieval Arabic nations but only came into widespread use in Europe in the latter part of the 16th century when serious efforts were made to improve practical methods of computation. François Viète, the leading French mathematician of the day, advocated use of decimals in 1579 but by this date they were not a new innovation, being routinely used by professional mathematicians.[28] However, they remained something of a mystery to the wider public until thoroughly explained by Simon Stevin in his little book *De thiende* (The tenth) published in Leyden in 1585.[29]

There is a price to be paid all the same for this commitment to decimal forms. In normal base ten arithmetic we exploit the fact that any number can be written as a sum of multiples of powers of ten. When expressing a fraction as a decimal, we are attempting to write the number as a sum of powers of $\frac{1}{10} = 0.1$. Unfortunately, even for very simple fractions such as $\frac{1}{3}$, this cannot be done, and the decimal expansion goes on without end: $\frac{1}{3} = 0.333 \ldots$. Certainly the ancients would have had none of this, as we have replaced a simple exact idea, a unit fraction, by a complicated idea involving a process of endless calculation. In practice however, we

27 The quickest way to decide is to cross-multiply: $\frac{19}{24} > \frac{47}{60}$ because $19 \times 60 = 1140 > 1128 = 24 \times 47$.

28 Al-Kashi (ca. 1436) of Samarkand considered himself the founder of decimal fractions. He certainly made great use of both decimal and sexadecimal (base 60) calculations although he may have been introduced to the practice from Chinese sources. However there is record of decimal usage in Arabia as far back as the 10th century.

29 Stevin undertook to explain 'how to perform with an ease, unheard of, all computations necessary between men by integers without fractions.'

appreciate that by truncating the decimal expansion after a certain number of places, (depending on what accuracy we demand) we can get by with a terminating decimal that approximates the exact fraction. As long as all the work is done by a calculator, everything goes smoothly enough. Any innaccuracy is trivial in comparison with the convenience of carrying out all our number work in the standard base ten frame of reference. Decimal expansions can be thought of as the closest we can get to having a single common denominator for all fractions.

It is natural to ask though, which fractions will have terminating expansions (and which will not)? The answer is, not very many. More often than not, the decimal expansion of a fraction goes into a recurring pattern: $\frac{3}{22} = 0.1363636\ldots$ with the 36 part repeating forever. Every fraction generates a recurring decimal in this way, and the length of the recurring block is no longer than one less than the value of the denominator. This can be seen by considering what happens when we carry out the corresponding long division sum: if the denominator is n, then the remainder after each step in the division takes on one of the values $0, 1, \cdots, n-1$. If at some stage the remainder is 0, the division terminates and so does the decimal expansion: for example, $\frac{3}{8}$ is exactly equal to 0.375. Otherwise the division continues forever but once a remainder is repeated, which is inevitable,[30] we shall be forced into the same cycle of divisions once more, thus giving us a recurring pattern whose block can be no longer than $n-1$. The expansion will terminate exactly when the denominator is a product of the prime factors 2 and 5 of our base 10 but not if there is any other factor

30 Here we are using the *Pigeonhole Principle*, if there are more than n envelopes to place into n pigeonholes then some slot will have more than one letter, that is some slot will be repeated.*

involved. For example, fractions with denominators 16, 40, and 50 are terminating but fractions like 1/14 and 1/15 will not terminate because the respective prime factors of 7 and 3 of the denominators spoil the party.

This does show however that whether a fraction's expansion terminates or not is not truly intrinsic to the number itself, but rather depends on the relationship of the number to the base you are using for your expansions. If for instance we worked in *ternary* (base three) then 0.1 would represent 1/3, as the 1 after the decimal point would stand for 1/3, and not 1/10, the way it does in decimal expansions.

The reverse process of turning a recurring decimal back into a vulgar fraction is also quite simple,* showing that there is a one-to-one correspondence between fractions and recurring decimals, and we can use which ever representation best suits our current purpose.

Does the class of fractions provide us with all the numbers we could ever need? The collection of all fractions, together with their negatives, form the set of numbers known as the *rationals*, that is all numbers that result from whole numbers and the ratios between them. They are adequate for arithmetic in that any sum involving the four basic arithmetic operations of addition, subtraction, multiplication, and division will never take you outside the world of rational numbers. If we are happy with that, this set of numbers, denoted by **Q**, is all we need.

There are visible hints however that we might go further if we wished. The modern outlook is firmly one in which we identify numbers with their decimal expansions and, for rational numbers, these are always recurring expansions. It is easy to imagine however arbitrary decimal expansions that do not have this or any other pattern. What is more, it is not hard actually to specify some such numbers. We cannot write down an infinite decimal expansion of course, but we could specify it by a particular pattern. As long as this was not a simple recurring pattern, we would be talking about a number that was not rational. For instance, what about the number $a = 0.101001000100001000001$ ... where the number of zeros between successive 1's increases by one in each block. This is not the expansion of a rational number in base ten (or in any other base). It seems it is very easy to find *irrational numbers*, ones that cannot be expressed as vulgar fractions.*

We might satisfy ourselves with the thought that we do not really need these numbers, as rationals form an adequate and self-contained world. This however is a very fragile view that shatters as soon as we try to apply measurement to geometry.

We all have heard about the number π, the ratio of the circumference of a circle to its diameter. If you ask a calculator the value of this number, it answers 3.1415927 .... There is no hint of recurrence here. Be that as it may, how can we tell? The length of the recurring block might be thousands of digits long, or perhaps the recurring pattern may not kick in till after millions of decimal places. In the same fashion, if you ask your calculator for the value of $\sqrt{2}$ it will give the answer 1.4142136 ..., and once again we are left wondering. In this case also our number might be rational, but we have no way of knowing.

Pythagoras however, knew, at least about the irrational nature of $\sqrt{2}$.[31] The Greeks did not think in terms of decimal expansions but were happy to recognize a length constructed in the geometry

---

31 The now familiar $\sqrt{}$ symbol is of course not Greek but was introduced in 1525 by Christoff Rudolff: it is meant to be reminiscent of the appearance of the letter r, standing for root.

of straightedge and compasses as representing a real quantity. In particular, Pythagoras's Theorem tells us that the longer side of a right-angled triangle whose shorter sides are each of unit length, is exactly equal to the square root of 2. (Actually you don't need Pythagoras's Theorem to see this, a certain ancient Indian argument suffices.*) Just as in the case of π, there is no recurring pattern to be seen. You might try to take square roots of a few other numbers. You will find that, unless you begin with is a perfect square in the first place, 1, 4, 9, 16, etc, the decimal display given for the answer will offer no encouragement to the view that everything should be rational in the end.

Pythagoras was able to prove that the square root of 2 was not equal to any known fraction, thereby showing that irrational numbers are real. In particular, you cannot measure the diagonal of a square with the same units with which you measure the side. They are fundamentally incompatible, or *incommensurable* as they are described in the classical texts. The story is the same for π, which is approximately equal to the fraction 22/7, but is different from it, and from any fraction that you might suggest. Although it is very difficult to prove this, the question for the square root of 2 can be settled easily enough by a simple contradiction argument. First we note that for any number c, the highest power of 2 that is a factor of c^2, is twice the highest power of 2 that is a factor of c, and so in particular the highest power of 2 that divides any square must itself be an even number. For example, $24 = 2^3 \times 3$ while $576 = 24^2 = 2^6 \times 3^2$, and in this case the highest power of 2 dividing the number does indeed double from 3 to 6 when we take the square. This is always the case and indeed applies not only to powers of 2 but to any prime factor of the original number.

Suppose now that $\sqrt{2}$ were equal to the fraction a/b. Squaring both sides of this equation allows us to deduce that $a^2 = 2b^2$. By the previous observation, the highest power of 2 that divides the left hand side of this little equation is even, while the highest power that divides the right hand side is odd (because of the presence of the extra 2). This shows this equation to be nonsense, and so it must not be possible to write $\sqrt{2}$ as a fraction in the first place. Like Pythagoras, we come face-to-face with the irrationals.

Arguments along these lines allow us to show that quite generally, when we take the square root, (or indeed the cube or higher roots) of a number the answer, if not a whole number, is always irrational, thus explaining why the decimal displays on your calculator never show a recurring pattern when asked to calculate such a root.*

Pythagoras, much to his own discomfort, had discovered that, in order to do his mathematics, he required a wider field of numbers than just the rationals. The Greeks regarded a number to be 'real' if its length could be constructed from a standard unit interval using only a straightedge (not a marked ruler, just an edge) and compasses. It turns out that although the square root operation does introduce irrationals, the full collection itself does not go very far beyond the rational. The set of *euclidean numbers*, as we shall refer to them, are all those that can be arrived at from the number 1 through carrying out the four operations of arithmetic and the taking of square roots, any number of times. For example, the $\sqrt{5 - \sqrt{3/2}}$ is a number of this kind. Even cube roots are beyond the grasp of the euclidean tools. This was the basis of perhaps the first great unsolved problem in mathematics. The first of three Delian Problems as they were known was the call to construct the cube root of 2, using only straightedge and compasses. Legend has it that this was the task set by the god in order to banish the plague from Athens, put in the form of exactly doubling the volume of an altar that was a perfect cube.

This problem remained untouchable in classical times—the Greeks never discovered the truth and bequeathed the puzzle to distant posterity. They did learn for instance that it was possible to construct this length using other mechanical tools, such as carpenter's tee and set square—indeed one particular construction is attributed to Plato himself. However the basic euclidean tools were regarded as pre-eminent, and until it was established that they were genuinely inadequate, the challenge set by the god lay unresolved. That the cube root of 2 lies outside the range of the euclidean tools was only settled in the 19th century, as it requires a precise algebraic description of what is possible using the classical tools in order to see that the cube root of 2 is a number of a fundamentally different type. It does indeed come down to showing that you can never manufacture a cube root out of square roots and rationals. When put that way, the impossibility sounds more plausible. However, that in no way constitutes a proof.

chapter 7

Glimpses of Infinity

The Greeks were left with an ambivalent attitude towards numbers. They understood the need to go beyond rationals, but were reluctant to go much beyond square roots as they were identified with lengths in euclidean geometry. At the same time they hankered after an understanding of cube roots, which seemed to reside at another level in the hierarchy that they were reluctant to recognize, for they could find no satisfactory way of getting to grips with it. There was also the vexed question of the value of π.

Archimedes ($c287$BC–$c212$ BC) had established that the area of a circle of radius r is πr^2. You will remember that π was defined as the ratio of the diameter of a circle to its circumference and, as such, there is no obvious reason why it should also be connected with its area. However Archimedes, in the 3rd century BC, had shown that a circle and triangle share the property that their areas are given by the formula of half base times height: if we interpret the 'base' of a circle as its circumference, and its 'height' as

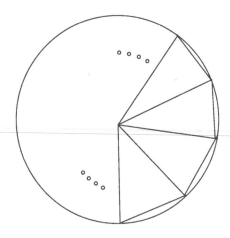

Figure 7.1. Approximating a circle by triangles on its circumference.

its radius, the triangle formula for area then takes on the form $\frac{1}{2} \times (2\pi r) \times r = \pi r^2$.

This is not a fluke, but comes about through approximating the area of a circle by a series of identical triangles within the circle, all with a common vertex at the center and base lying on the circumference (see Fig. 7.1).

Using this approach, Archimedes went on to show that π lies between the rational numbers $3\frac{10}{71} = 3.1408$ and $3\frac{1}{7} = 3.1428$. Indeed, another of the classical Delian Problems was the challenge to Square the Circle—that is to say construct, using only the euclidean tools, a square with the same area as a given circle. This is also impossible due to the nature of π, for it is a number that is not rational nor indeed euclidean. For example, if we could construct a square with the same area as a circle of radius one unit then the side length of that square would be $\sqrt{\pi}$. However, if this length could be constructed then so could the square of this length, which is π itself. The Greeks could not have had any true inkling, but π lies outside the range of any number type they were considering, as no expression for it can be written using fractions, even if we allow ourselves free use of the operations of cube or higher roots. This fact was only finally proved in the 1880's by Lindermann.

In contrast, in the mind of a modern person, the idea of number is not so bound up with constructible geometric lengths, which is a corner of school geometry that largely lies neglected and almost forgotten. To us, who have been brought up on decimals, it seems natural to call any number 'real' as long as it has a decimal expansion of any kind, even if it is infinite and lacks apparent pattern. This easy going attitude brings with it much more than we might have bargained for. Free use of decimals has many benefits but also opens the door to a host of questions concerning the infinite.

It was Galileo who was the first to alert us to the fact that the nature of infinite collections is fundamentally different to that of finite ones. As mentioned early in our discussion, the size of a *finite* set is smaller than that of a second set if the first can be put into a one-to-one correspondence with just a portion of the second. However, infinite sets by constrast can be made to correspond in this way to subsets of themselves (whereby the term *subset* I mean a set within the set itself that forms part of the original). We need go no further than the sequence of counting numbers to see this, $1, 2, \cdots$. It is easy to describe any number of subsets of this collection, which themselves form an infinite list, and so are in one-to-one correspondence with the full set: the odd numbers, $1, 3, 5, 7 \cdots$, the square numbers, $1, 4, 9, 16 \cdots$ and, less obviously, the prime numbers, $2, 3, 5, 7 \cdots$. Indeed as Galileo observed, it is just this property that defines the different character of infinite collections as compared to finite ones: if you take away some members of a finite collection, then the set you are left with certainly has fewer members, but this is not necessarily the case when dealing with infinite collections.

The Hilbert Hotel

This rather extraordinary hotel is always associated with the mathematician David Hilbert (1862–1943) and it serves to bring to life the strange nature of the infinite. Out in the depths of space there floats Hilbert's Hotel, the biggest in the universe. In fact it has infinitely many rooms, one for each counting number, and a sign that boasts that there is always room at the *Hilbert Hotel*.

However one night, it is in fact full and, much to the dismay of the desk clerk, one more customer fronts up demanding a room. When the clerk (who had not had a course in Infinite Hotel Management) apologetically admits that the hotel is full, the customer, having traveled thousands of light years, is understandably irate and points to the claim on the advertising hoarding. However, an ugly scene is avoided when the manager intervenes and takes the clerk aside to explain how to deal with the situation. We simply act as follows, says the manager. We tell the occupant of Room 1 to move to Room 2, that of Room 2 to move into Room 3 and so on. That is to say we issue a global request that the customer in Room

n should shift into room $n + 1$, and this will leave Room 1 empty for this gentleman!

And so you see there *is* always room at the Hilbert Hotel. But how much room?

The next evening the clerk is confronted with a similar but more testing situation. This time a spaceship with 42 passengers arrives, all demanding a room in the already fully occupied hotel. The clerk has however learnt his lesson from the previous night and sees at once how to extend the idea to cope with this additional group. He tells the person in Room 1 to go to Room 43, that of Room 2 to shift to Room 44, and so on, issuing the global request that the customer in Room n should move into Room number $n + 42$. This is a bit trickier but it does leave Rooms 1 through to 42 free for the new arrivals and our clerk is justly proud of himself for dealing with this new version of last night's problem all by himself.

The final night however the clerk again faces the same situation—a full hotel but this time, to his horror, not just a few extra customers show up but an infinite space coach with infinitely many passengers, one for each of the counting numbers 1, 2 etc. The overwhelmed clerk tells the coach driver that the Hotel is full and there is no conceivable way of dealing with this lot. He might be able to squeeze in one or two more, any finite number perhaps, but not infinitely many more. No way!

An infinite riot might have ensued except again for the timely intervention of the manager, who informs the coach driver that there is no problem at all. There is always room at Hilbert's Hotel for anyone and everyone. He takes his panicking desk clerk aside for another lesson. All we need do is this, he says. We tell the occupant of Room 1 to shift into Room 2, that in Room 2 to shift to Room 4, that in Room 3 to go to Room 6, and so on. In general the global instruction is that the occupant of Room n should move into Room $2n$. This will leave all the odd numbered rooms empty for the passengers of the Infinite Space Coach. No problem at all!

The manager seems to have it all under control. However even he would be caught out if a spaceship arrived that somehow had the technology to have one passenger for each point in the continuum of the real line. One person for every decimal number would totally overrun Hilbert's Hotel, and we shall see why in the next section.

Cantor's Comparisons

All this may be surprising the first time you think about it, but it is not difficult to accept that the behavior of infinite sets may differ in some respects from finite ones, and this property of having the same size as one of its subsets is therefore a case in point. In the 19th century however Georg Cantor (1845–1918) went much further and discovered that not all infinite sets can be regarded as equal either. This revelation was truly unexpected and novel in nature. It is however not hard to appreciate once your attention is drawn to it.

Cantor asks us to think about the following. Suppose we have any infinite list L of numbers $a_1, a_2, \cdots$ thought of as being given by their decimal expansions. Then it is possible to write down another number, a, that does not appear anywhere in the list L: we simply take a to be different from a_1 in the first place after the decimal point, different from a_2 in the second decimal place, different from a_3 in the third decimal place, and so on—in this way we may build our number a making sure it is not equal to any number in the list. This observation looks innocuous but it has the immediate consequence that it is *absolutely impossible* for the list L to contain *all* numbers, because it does not contain the number a. It follows that the set of all real numbers, that is all decimal expansions, cannot be written in a list, or in other words *cannot* be put into a one-to-one correspondence with the natural counting numbers. The collection of all real numbers is therefore larger in a sense than the collection of all positive integers. Even though both are infinite, the sets cannot be paired off together the way the even numbers can be paired with the list of all counting numbers. In fact Cantor's Diagonal Argument as it is called can be applied to all the numbers in the interval 0 to 1, with the same conclusion, as we may build our number a in those circumstances to also lie in this range. I mention this as we shall make use of that fact shortly.

Cantor's result is rendered all the more striking by the fact that many other sets of numbers can be put into an infinite list, including the Greeks' euclidean numbers. A little ingenuity is involved, but once a couple of tricks are learnt, it is not hard to show many sets of numbers are *countable,* which is the term we use to mean that the set can be listed in the same fashion as the counting numbers. Otherwise a set is called *uncountable.*

For example, let us take the set of all integers Z, which comes to us naturally as a kind of doubly infinite list. We can however rearrange it into a row with a starting point: $Z = \{0, 1, -1, 2, -2, 3, -3, \cdots\}$, by pairing each positive integer with its opposite we create a list where every integer appears—none will escape. We can also do the same with the rationals: start with 0, then list all the rationals that can be written using all integers no more than 1, (which are 1 and -1) then those that involve no number higher than 2, (which are $2, -2, \frac{1}{2}, -\frac{1}{2}$) then those that only use numbers up to 3, and so on. In this way the fractions, (positive, negative, and zero) can be arranged in a sequence in which they are all present and accounted for. The rationals therefore also form a countable set, as do the euclidean numbers, and indeed if we consider the set of all numbers that arise from the rationals through taking roots of any order, the collection produced is still countable. We can even go beyond this: the collection of all *algebraic numbers,* which are those that are solutions of ordinary polynomial equations* form a collection that can, in principle, be arrayed in an infinite list: that is to say it is possible, with a little more crafty argument, to describe a systematic listing that sweeps them all out.

What we have allowed to happen in casually accepting any decimal expansion is to open the door to what are known as the *transcendental numbers*, those numbers that lie beyond those that arise through euclidean geometry and ordinary algebraic equations. Cantor's argument shows us that transcendental numbers exist and, in addition, there must be infinitely many of them, for if they formed only a finite collection, they could be placed in front of our list of algebraic numbers (the non-transcendentals), so yielding a listing of all real numbers, which we know is impossible. What is striking is that we have discovered the existence of these strange numbers without identifying a single one of them! Their existence was revealed simply through comparing certain infinite collections to one another. The transcendentals are the numbers that fill the huge void between the more familiar algebraic numbers and the collection of all decimal expansions: to use an astronomical comparison, the transcendentals are the dark matter of the number world.

In passing from the rationals to the reals we are moving from one set to another of *higher cardinality* as mathematicians put it.

Two sets have the same *cardinal number* if their members can be paired off, one against the other.* What can be shown using the Cantor argument is that any set has a smaller cardinal number than the set formed by taking all of its subsets. This is obvious for finite collections: if a set has three elements, a, b, and c, there are eight subsets that we can form from these three elements: three consisting of just one of the three elements, three pairs $\{a, b\}$, $\{b, c\}$, and $\{a, c\}$ (the order in which the letters are listed is not important), and we don't forget the original set itself and the empty set (the set with no members). This gives us the eight possibilities. In general if we begin with a set of n members there are 2^n subsets that can be formed in this way. (See Note 10, Chapter 13.) What about the infinite collection of counting numbers, $\{1, 2, 3, \cdots\}$? Here is where Cantor's argument once again comes into its own and shows that even for infinite sets, the set of subsets is always strictly larger than the original set.*

There is another approach to the size of infinite collections when the sets in question are regarded as being ordered, and this provides a different source of comparison. But more of this later in the chapter.

Returning to the transcendental numbers, some readers may yet be surprised that they have never come across them before. This however is not so unexpected as, by their very nature, they are the numbers that do *not* arise through the ordinary calculations of arithmetic and the extraction of roots, which represent the family of basic arithmetical operations. What is more, the transcendentals form a very secretive society, and those in it do not readily admit to membership of the club. For example, the number π is an example of a transcendental but this is not a fact that it openly reveals.

Perhaps an even more important instance of a transcendental is the number $e = 2.71828\ldots$. This number arises constantly in the calculus: it is the base of the so-called *natural logarithm*, the function that tells you the area under the graph of the reciprocal function. It is also the limiting value of the sequence of numbers you get when you raise the ratio of two consecutive integers, $\frac{n+1}{n}$, to the power n. (Ask your calculator for the value of $(21/20)^{20}$.)

This sequence arises when we consider the problem of the limiting value of an interest rate as you make the interval of repayment shorter and shorter from annually, to monthly, to daily, to every second, and so on. In particular, suppose that you invest one unit in a scheme that promises to double your money every year, that is to say pays interest at an annual rate of 100%. After one year you will have 2 units. You would be better off however in a scheme that paid 50% interest every six months for then you could re-invest the interest paid half way through the year and earn interest on that interest in the second half year. Every six months your capital would be multiplied by a factor of $1\frac{1}{2}$ or, to put it another way, at year's end your account would hold $(1 + \frac{1}{2})^2 = 2.25$ units, an effective APR of 125%. Better yet would be an account that accrued interest monthly—your nest-egg would be multiplied by $1\frac{1}{12}$ each month, yielding: $(1 + \frac{1}{12})^{12} = 2.613$ units, an annual percentage rate of 161.3%. The shorter the waiting period for the next interest payment, the better for the investor, so that if your account earned interest daily, you would be better off still. If we take this to the limit, we would have an account that accrues interest continuously. This does not however break the bank for the following reason.

The general situation is that interest is paid n times per year which means that you initial investment is multiplied by the factor

$(1 + \frac{1}{n})$, n times in all. The limiting multiplier therefore that would apply in the continuous interest case is the limiting value, as n increases without bound, of the number

$$\left(1 + \frac{1}{n}\right)^n = e = 2.71828 \ldots.$$

The limiting interest rate is 171.82...%, and is not infinite!

Yet another way in which the mysterious e arises, is through the sum of the reciprocals of the factorials, and this gives a way of calculating e to a high degree of accuracy:

$$e = 1 + \frac{1}{2!} + \frac{1}{3!} + \frac{1}{4!} + \cdots$$

This avenue of investigation also allows you to show, relatively easily, that e is an irrational number.* Showing that it is not just irrational, but transcendental, requires quite a bit more work.

Since e crops up in a variety of distinct and fairly simple ways, it persistently appears throughout mathematics, often where you would not expect to meet it. For example, take two packs of playing cards, turn over the top card of each deck, and compare. Continue doing this until you have exhausted the packs. What are the chances that, at some stage, there is a perfect match? That is to say, on one turn or another the cards showing are exactly the same, be it the eight of clubs, Queen of Hearts, or whatever. It works out that the proportion of times this experiment yields at least one such match is as near as makes no difference to $1/e$, that is about 36.8%.* Apparently this is quite a bit higher than most people would guess, and so this game forms a good basis for a 'bar room bet' as punters are likely to grant you odds of 5/1 or better against a coincidence turning up.

The truly special status of *e* is undeniable in a way that of π is perhaps not. After all, why should we assign a special symbol to π? The answer is, because π is the ratio of the circumference to the diameter of a circle. But since the radius is used more than the diameter, should we not afford a special status to the number whose value is 2π, rather than π?[32] Indeed many mathematicians would have preferred to grant a particular symbol to the number $\pi/2$ instead, as it occurs more often that does π in mathematical calculations. This may be because, as we travel around the circumference of a circle of unit radius $\pi/2$ units, we trace one quarter of the circle, corresponding to turning through one complete right angle and, as Pythagoras showed, the right-angle is the most fundamental of geometric ideas.

Transcendental numbers then are numerous but exceedingly slippery. As a rule of thumb, a number that arises in mathematics is almost always transcendental unless it is obvious that it is not. However, showing that a particular number *is* transcendental can be exceedingly difficult. Number theory throws up endless problems of this kind where everyone feels sure what the answer must be but at the same time no-one has any real idea how it could ever by proved.

Structure of the Number Line

All this can be recapped in the language of simple equations. The rational numbers, which form a countable set, are the numbers

32 The 15th century Arabic mathematician Al-Kashi calculated 2π correct to 16 decimal places.

that arise as solutions of simple linear equations: the fraction b/a is the solution to the equation $ax - b = 0$ (a and b are integers). The numbers like $\sqrt{2}$ that do not arise in this way are called irrational, and they form an uncountable collection that cannot be paired off with the counting numbers in the way that the rationals can. Within the set of irrationals there are the transcendentals, which are the numbers that never arise as the solutions of equations of these kinds even if we allow higher powers of x. It is known that π is an example of a transcendental number, but $\sqrt{2}$ is not, as it solves the equation $x^2 - 2 = 0$. All the same, the transcendental numbers comprise an uncountable set as well.

There are however completely different ways of looking at the size of infinite sets of numbers that are revealed if we look at the distribution of the various number types that knit together to bind the number line into a continuum. The rationals may only be a countable collection of numbers but the collection is densely packed within the line in a way that the integers plainly are not. Given any distinct numbers, a and b, there is a rational number that separates them. The average of the two numbers, $c = \frac{a+b}{2}$, certainly is a number lying between them, but it may be irrational. However, if c is irrational we can approximate it by a rational number d, with a terminating decimal expansion, by letting d have the same decimal representation as c up to a very large number of decimal places. For example, if we take $\sqrt{2} = 1.414 \ldots$ we have that $\sqrt{2}$ differs from 1.414 by less than 0.001 and each time we take another decimal place we guarantee finding a rational number that approximates $\sqrt{2}$ more accurately (on average, ten times more accurately) than the previous one. If the number of initial places in which they agree is sufficiently large, then their difference will be so small that both c and d will lie between a and b. The number of

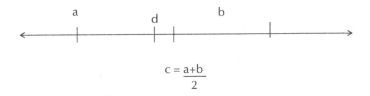

Figure 7.2. Locating a rational between two given numbers.

places we need to take after the decimal place will depend on just how close *a* and *b* are to each other to begin with, but it is always possible to find a rational *d* that does the job (see Fig. 7.2). We say that the set of rational numbers is *dense* in the number line for just this reason. Of course we can, by the same argument, show that there is another rational, splitting the interval from *a* to *d* say and, in this way, we are led to the conclusion that infinitely many rational numbers lie between any two numbers, however small the difference between these two numbers might happen to be. In particular there is no such thing as the smallest positive fraction, for, given any positive number there is always a rational lying between it and zero.

Not to be outdone, the set of irrationals also forms a dense set. Before explaining this, I point out that once we have identified one irrational, the Pythagorean number $\sqrt{2}$ for example, the floodgates open and we can immediately identify infinitely more. When we add a rational to an irrational the result is always an irrational.* For example, $\sqrt{2} + 7$ is irrational by dint of this reasoning. In a similar fashion, if we multiply an irrational number by a rational number (other than 0) the result is another irrational number. In particular we can find an irrational number of size as small as we like: $t = \frac{\sqrt{2}}{n}$ is irrational for any counting number n and by taking n larger and larger we can make t as close to 0 as we please.* As with the rationals we therefore see that there is no smallest positive irrational, and hence there is no such thing as the smallest positive number.

Returning to our given numbers, a and b, once again let c be their average. If c is irrational, we have a number of the required kind. If on the other hand c is rational, put $d = c + t$, where t is the irrational number of the previous paragraph. By what has gone before, d will also be irrational, and if we take n large enough, we can always ensure that d is so close to the average c of the two given numbers a and b that it lies between them. In this way we see that the irrational numbers too form a dense set and, as with the rational numbers, we can infer that there are infinitely many irrational numbers lying between any two numbers on the number line.

And so the set of rationals and its complementary set of irrationals are in one way comparable (they are both dense in the number line) and in another not (the first set is countable, the second not). The question of how we should measure the size of these sets is then not totally resolved by the above discussion, and this has had important ramifications in the theory of probability, which is a major application of the mathematics of numbers. Gambling problems, like one-off card games and lottery draws, can be dealt with perfectly adequately using rational numbers. Although such problems of chance can be very tricky and involve subtle techniques, there is no real difficulty in interpreting the results of the calculations. However, when the infinite began to enter into problems of probability, real difficulties emerged about how to proceed. Certain problems involving infinite sets led to different answers depending on just how you looked at them. For that reason, a theory of general probability took a long time to mature and fully establish itself. Even problems involving games that could carry on indefinitely led to confusion. Although probability theory is now one of the pillars of scientific thinking, for a long time it lacked respectability. Absence of a sound framework in which to operate left mathematicians only able to deal with a limited range of problems involving chance.

The general setting in which probability theory now sits is within the area of mathematics known as *Measure Theory*. This topic allows us, for instance, to measure the size of the set of rational numbers in the unit interval from 0 to 1. The interval itself has measure 1, as we would expect. Any countable set like the rationals has measure zero. The complementary set of irrationals then has measure $1 - 0 = 1$ also. However, although the countable sets are all of measure zero, the uncountable subsets can have very different characters and display measures of any value between these two extremes. For example, an interval on the real line of length l, has measure l, as you would expect— indeed if measure theory assigned a different value it would be unworthy of its name. The surprising aspect is that it is possible for uncountable sets also to have measure zero, showing that knowing that a set is uncountable tells us nothing whatsoever about its measure. The standard example of an uncountable set of measure zero is Cantor's Middle Third Set, a kind of fractal pattern. It comes about by removing the middle third of an interval, and then continuing with this process on the intervals that remain. Surprisingly perhaps, there are some points, indeed very many points, which survive this infinite cull and together they form a set of great interest that we shall meet in Chapter 11 when we re-examine the number line with the aid of an infinitely powerful microscope.

Infinity Plus One

A different kind of numbering, also invented by Cantor, comes to light when we start to look at infinite sets that arise with a natural order. We motivate this through an example. One aspect of the infinite that is surprising the first time you see it is that it is possible to add together an infinite sequence of positive numbers but never get past a finite limit. The standard example is the series

$$\frac{1}{2} + \frac{1}{4} + \frac{1}{8} + \frac{1}{16} + \cdots$$

The limit of this series is 1, meaning that as you add up initial sequences of numbers from the series the sum draws ever nearer to 1, without reaching it—we say that 1 is the limit of this series, as it is the smallest number that is not reached, nor exceeded, by summing the terms. The reason for this is that after summing an initial sequence of terms from the list (we call such a *partial sum*), the next term you add equals half the distance to your final target of 1.

If we list all the partial sums we could meet along the way together with the limit we have an infinite set S of numbers of a somewhat peculiar form:

$$S : \frac{1}{2} < \frac{3}{4} < \frac{7}{8} < \frac{15}{16} < \cdots < 1.$$

It is quite natural to consider the number of this ordered set to be infinity plus one, as there is an infinite ascending chain of numbers, followed by a single number that exceeds them all. This is how the *ordinal numbers* are introduced. The finite ordinal numbers are just the ordinary counting numbers considered in the usual order, $1 < 2 < 3 < \cdots < n < n + 1 < \cdots$. *After* all of these there is the first infinite ordinal, the ordinal number of the *entire set* of counting numbers, denoted by ω. If, as in our example above, we then have an additional member that lies above them all, we obtain a set ordered in the fashion of our set S above, the ordinal number of which is $\omega + 1$. If we then extend S, say by including 2 which lies above 1, the ordinal number of the set that results would be $\omega + 2$ and so on, $\omega + 3$, $\omega + 4$, $\cdots$.

It makes a difference however to the nature of the ordering whether we place the new element at the beginning or at the end of the set, and this is reflected in the ordinal number of the set. For instance, the ordinal of the counting numbers is ω. If we extend the counting numbers to include 0, the ordered set we obtain is of course

$$0 < 1 < 2 < \cdots < n < \cdots$$

which still has ordinal type ω for, as an ordered set, it has the same structure as the counting numbers beginning from 1. It certainly is not of the same ilk as the set S above of type $\omega + 1$, as there is no member of the set that stands supreme above them all—that is to say the set lacks a greatest element.

The lesson to be learned from all this is that when it comes to the addition of the infinite ordinals, order does matter: $1 + \omega = \omega < \omega + 1$.

We could go further, the set

$$0.9 < 0.99 < 0.999 < \cdots < 1 < 1.9 < 1.99 < 1.999 < \cdots$$

consists of one set of order type ω, followed by another and so has order type: $\omega + \omega$, which we denote by $\omega \times 2$ or $\omega 2$, for we are taking copies of sets of ordinal type ω and doubling up. If on the other hand we consider 2ω, which we interpret as ω copies of ordered pairs, the set that results still has order type ω, for if the ordered pairs are denoted as $(1, 1'), (2, 2'), \cdots, (n, n'), \cdots$ then the ordered set represented by 2ω has order

$$1 < 1' < 2 < 2' < 3 < 3' < \cdots < n < n' < \cdots$$

which still has the order type of the counting numbers, which is ω.

However we can always move on to larger ordinals by adding new elements above old ones. If we keep going we get:

$$\omega 2 < \omega 2 + 1 < \omega 2 + 2 < \cdots < \omega 3 < \cdots < \omega 4 < \cdots < \omega^2$$

The ordinal ω^2 is the ordinal type of a set that corresponds to an infinite array:

$$a_1 < a_2 < \cdots < a_n < \cdots$$
$$b_1 < b_2 < \cdots < b_n < \cdots$$
$$c_1 < c_2 < \cdots < c_n < \cdots$$
$$\vdots$$

where all the a's precede all the b's, which come before all the c's, and so on. After this comes,

$$\omega^2 + 1 < \omega^2 + 2 < \cdots < \omega^2 + \omega 2 < \omega^2 + \omega 2 + 1 < \cdots$$
$$< \omega^2 + \omega 3 < \cdots < \omega^2 2 < \omega^2 2 + 1 < \cdots < \omega^3 < \cdots$$

And we can continue as long as we fancy:

$$\omega^4 < \cdots < \omega^5 < \cdots < \omega^\omega < \omega^\omega + 1 < \cdots < \omega^{\omega 2}$$
$$< \cdots < \omega^{\omega^\omega} < \cdots < \omega^{\omega^{\cdot^{\cdot}}}$$

where there are ω of the ω's in the last power tower. This ordinal number is known as ε_0, and it is the first ordinal that you cannot reach from lower ordinals through a finite number of addition, multiplication, and exponentiation operations. And then comes $\varepsilon_0 + 1 < \cdots$.

As you will already appreciate, ordinal arithmetic has a unique flavour all its own, and for that reason it is a rich source of strange and exotic examples in various areas of mathematics, particularly topology, the study of space at its most abstract. Moreover Cantor's ordinal numbers have been developed further quite recently. In order to generalise the way that the ordinary real numbers fill in all the gaps between the integers, the British mathematician John Conway has invented what he calls the *surreal numbers,* the purpose of which is to fill the gaps between Cantor's ordinals.

chapter 8

Applications of Number: Chance

Probability is one area that has taken enormous strides, not only in theoretical development, but in winning recognition of its importance. On the one hand it is found in advanced theoretical physics and economics, and on the other it has percolated down into the beginning of school study. Up until the latter part of the 18th century, it was not properly recognized as an area of applied mathematics. Although chance and games of chance have been with us for millenia, and despite the fact that numbers are clearly involved, the subject was never treated as ripe for thorough investigation. Perhaps, in the eyes of scholars, probability was sullied by its association with gambling, making it unfit for serious study. At the same time the random nature of chance may have suggested that it is the very opposite of mathematics, which traditionally was seen as the the subject of eternal and rigid truths. Be that as it may, probability has turned out to be one of the most fruitful and active fields for mathematical analysis and continues to produce surprises.

At its simplest conceptual level, probability deals with a finite collection of equally likely outcomes. The probability of an event is then a simple fraction, between 0 and 1 inclusive, representing the proportion of favourable outcomes. For example, when an ordinary die is tossed, the probability of rolling a 5 or a 6 is $\frac{2}{6} = \frac{1}{3}$, as there are 6 outcomes, 2 of which result in the sought for event. This is taken to mean that if we were to conduct this experiment a very large number of times and record the proportion of favourable cases, that ratio would, after a wobbly start, settle down very close to this fraction. Probablilities such as these can then be the basis of setting odds for various games of chance.

In this context of a limited number of equally likely outcomes, probability problems are counting questions: to find the answer we need to count the number of favourable outcomes, and divide this by the total number of outcomes possible. Problems involving a pair of dice require a little more thought. For instance:

What is the probability of rolling a total of 7 with a pair of dice?

Here we are interested in totals so from that viewpoint there are 11 possibilities because any total from a high of 12 (double sixes, get out of jail free) down to a low of 2 (snake eyes) is possible. The only outcome that we are regarding as favourable is a total of 7, so we might be tempted to jump to the conclusion that the answer is 1/11. A little experimentation with real dice would soon convince us that we have made a mistake: a total of 7 comes up more often than that because there are several combinations that yield a total of 7 (although only one combination yields each of 2 and 12). It is true that we have regarded the *event space* (the word space is used a lot in mathematics) as consisting of eleven events but since they are *not* equally likely this will not do. In order to apply the basic definition, due to the 18th century French mathematician Laplace, we need to re-interpret the space as consisting of equally likely events. This is done by listing the $6 \times 6 = 36$ possible outcomes that arise by listing the outcome of each die thrown in order: (4, 1) for instance would mean 4 showing on the first die and 1 on the second. Having done this, we count those ordered pairs that yield a total of 7, we find 6 of them in all, and so the answer is $\frac{6}{36} = \frac{1}{6}$; a pair of dice shows a total of 7, on average, one time in six.

Again these arguments are taught to high school students who are expected to master them but it was not so in centuries past. They represent part of a very modern outlook. In the 18th century, leading scientists sometimes indulged in wrong-headed notions about probabilities, even after the faults in their reasoning were revealed. The above type of error was made in even simpler situations such as tossing a pair of coins. The probability of a pair of heads is 1 in 4, but it was often argued that it was 1 in 3, as there were *three* possible outcomes: two heads, two tails, or one of each. The trouble is that this last event can arise in two distinct and equally likely ways (HT or TH) and so the events that comprise this trio are not equally likely. We have to imagine the event space to consist of the four equally likely outcomes of (HH, HT, TH, and TT) to start getting the answers right. Any learned gentlemen who stubbornly persisted in errors of this ilk would stand to lose a lot of money in age old gambling games like 'two-up'.

Although slow to develop, the modern origin of probability can be traced back to Blaise Pascal in the 17th century.[33] The

33 Girolamo Cardano (1501–1576) was a famous physician and mathematician of Milan who will feature later in our story. His main source of income however was gambling and he wrote *Book on Games of Chance*, which was published posthumously in 1663.

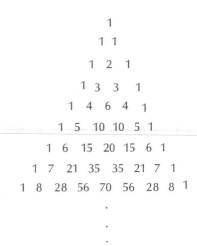

Figure 8.1. Pascal's triangle.

first serious question he tackled was a gambling problem set by his friend Chevalier de Mere who wanted to know how the chips should be divided among players in a game of dice that was interrupted before its completion. These dice problems led Pascal to discover the triangle of numbers that bears his name (see Fig. 8.1).

Each number in the body of the triangle is the sum of the two above it. The triangle, which can be continued indefinitely, gives the full list of choice numbers as we have called them, as was explained in Chapter 4. Even in the 17th century the triangle was more than 600 years old but Pascal's study revealed much that was new about it and about games of chance.

Pascal was a brilliant mathematician and philosopher, but suffered throughout his life from tortures of both the body and mind. He is noted for coming up with his 'wager', in which he argued that he was prepared to believe in God on the grounds that he had nothing to lose if he was wrong, and everything to gain should he be right. Even assuming that belief is a voluntary state of mind, this final conclusion is far from certain—it is easy to imagine that God might have a special Hell reserved for people who believed in him only as an insurance policy.

Some Examples

Next, a slightly tougher problem. In the old British Commonwealth, the main team game of bat and ball is not baseball but cricket, which is hotly contested on an international level by the nine so called Test Cricket nations, England, Australia, New Zealand, India, Pakistan, Sri Lanka, The West Indies, South Africa, and Zimbabwe. Countries outside the charmed circle are often totally unaware of the importance of these contests: cricket stars such as India's Sachin Tendulkar earn incomes rivalling those of any international soccer 'galactico' or American baseball player.

A test series in cricket can involve five matches each of up to five days duration. Before the match commences, the captains of the respective national sides toss a coin, and the winner chooses whether or not to bat first. This decision is crucial as the toss often confers on the winning captain a real advantage. It is surprising how frequently one captain loses all or nearly all the tosses and then goes on to lose the series. It can be very demoralising for a team to have a captain who is a loser but what are the chances?

What is the probability that one captain or the other loses all, or all but one of the five tosses?

To be definite, let us suppose that we are immersed in the traditional 'ashes' battle where England take on Australia. The outcome of all five tosses can be coded as a binary string of five symbols, each a 0 or a 1, which respectively stand for Australia and for England winning the toss. Since there are 2 possibilities for each of the 5 events, the total number of possible strings is $2^5 = 32$. There is therefore just 1 chance in 32 that the English captain loses every toss, as represented by the string 00000. However there are 5 strings where he loses all but one of the tosses, corresponding to the 5 places where the 1 occurs in the string. In total then, the probability that the English captain wins no more than one toss is $\frac{6}{32}$. By the same argument, this also represents the chance of the Australian captain winning at most once, so that the probability that one or other of the captains wins only once, or not at all, is $\frac{12}{32} = \frac{3}{8} = 37.5\%$. This means that we can expect in more than one series in three that one captain or the other will lose 4 or 5 times out of 5. When that happens, he may give the impression of being cursed with bad luck, but this is something of an illusion.

It is true that if a fair coin is tossed many thousands of times, the proportion of heads showing will approach the expected value of one half. However, although this is inevitable in the long run, at first it is quite likely that either H or T will build up quite a lead that may well persist for a surprisingly long period. The coin has no memory and if it does, by chance, give tails a lead, it feels under no obligation to peg that lead back—indeed it is just as likely to increase the lead of tails as it is to start balancing the account. This is a point that is not always appreciated. On the contrary, people often insist the opposite is true and appeal to what they call 'the law of averages' to justify the claim that luck tends to balance itself up. They are not entirely wrong, it is just that they expect fairer treatment to emerge more quickly than it often does. When bad luck persists, they feel themselves persecuted by fate in a way that they find hard to understand. On the other hand, a lucky streak may engender overconfidence, leading to recklessness that ends in tears. The numbers though explain how it comes about that chance is often pretty rough in dispensing equity and justice.

The next level of difficulty in probability matters is experienced with problems involving card hands: find the probability of being dealt a certain type of hand in poker and the like. The facet of the problem that is new is the sheer size of the numbers involved: the number of distinct 5-card hands is the number of ways of choosing 5 cards from a deck of 52, which equals $2, 598, 960$. However, all these choice numbers, or binomial coefficients as they are often known, can be expressed in terms of factorials. These numbers are very amenable to manipulation since they involve products of factorial numbers that often cancel down to very simple ratios. This allows these problems to be solved in practice by hand, without the use of calculators. For example, the probability of being dealt a flush in poker (all cards of the one suit) is $\frac{33}{16.660}$ which is just under 0.2%: about one chance in 500. This is a rare event, but not an impossible one: if the dealer deals himself a flush, you might have to believe it. But not four aces—there are only 48 hands with four aces so the probability of dealing that is $\frac{48}{2,598,960} = 0.0000185$, something less than one in fifty thousand.

Real life problems throw up all kinds of complications that are dealt with through a variety of techniques, some of which are diabolically clever. The following counting problem is of a type that arises in many guises, from abstract algebra to sub-atomic physics. Suppose that we have a party of eight teenagers who go to the movies and find three films on offer: *Boy Story, Revenge of Puff the Magic Dragon*, and *Titanic II*.

If eight people choose from three different films, how many different orders for the ticket purchases are possible?

Notice that, from the viewpoint of ticket sales, it does not matter *who* goes to see what film but only how many customers there are for each one. An approach that leads to the solution comes from thinking about what might be written down on a piece of paper by the member of the group whose task it is to buy the tickets for everyone in the party. She could write down two slash marks to separate the order into the three types available, heading the columns that result B, R, and T to remind her of what each means. As each person tells her their choice, she puts a cross in the corresponding column. Two possible orders are given below in Fig. 8.2: the first corresponds to two people wanting to watch *Boy Story*, with three each for the other two films, while the second order card

B	R	T		B	R	T
x x	x x x	x x x		x x x		x x x x x

Figure 8.2.

shows that no-one wants to watch *Puff*, with three customers for *Boy Story* and the rest for *Titanic II*.

What the buyer of the tickets has done is to devise a code by which every possible order of ticket purchase for the eight people is written as a string of ten symbols, consisting of eight crosses and two slashes, and so the answer to our question is equal to the number of ways that these ten symbols can be arranged.

Now an arrangement of the ten symbols is decided once we choose where to place the two slashes from among the ten positions possible, for then the crosses simply occupy all the remaining positions. (Note that the slashes are free to go anywhere: for example two slashes followed by eight crosses means they all want to see *Titanic II*.) There are ten choices for the position of the first slash and, for each such choice, there are nine positions left for the second slash, so there are $10 \times 9 = 90$ ways that she can write down one slash, and then another. However, we have to divide this answer by two, as for each choice of two positions for the slashes we can write either the one on the right, or on the left first, yielding the same outcome, so there are $90/2 = 45$ different ways the slashes can be placed in the row of ten positions. In other words there are 45 different orders possible for the collection of eight tickets.

Next we have a counting problem of a different kind. In any knock-out tennis tournament with n players there are $n - 1$ matches in all, as each match eliminates one player, leaving the undefeated champion triumphant at the conclusion. Two players are selected at random, that is to say their names are drawn from a hat.

What is the probability, on a scale from 0 to 1, that the given pair of players will play one another during the course of the tournament?

The answer is not hard to obtain once we convince ourselves to think of the *pairs* as individual units. Before we solve this problem though, let me remind you of how this kind of thing works through a simpler example.

Suppose there are 100 raffle tickets: you have bought one, and five are to be drawn out. The probability that you win a prize in the raffle is then 5/100, or one in twenty. The above situation is essentially the same. There are a certain number, m, of pairs (exactly how many we shall sort out in a moment), $n - 1$ of them will be chosen to play, and we ask for the chances of one particular random pair being selected. The answer to the question is therefore $\frac{n-1}{m}$.

How many possible pairs m are there? Each player, and there are n of them, can form a pair with $n - 1$ others. This gives the number $n(n - 1)$ which is *twice* the number of pairs as each pair of players, A and B, is counted twice in this way—once as A playing B and also as B playing A, and so, as in the previous question, we need to divide this number by 2 to get the total number of pairings possible: $m = \frac{n}{2}(n - 1)$. The answer to the question is then obtained by dividing $n - 1$ by this number m: the common factor of $n - 1$ in numerator and denominator cancels to leave the very simple expression, $\frac{2}{n}$, as our answer.

A surprisingly simple answer indeed that is easily verified for small numbers of players: if $n = 2$, the smallest number possible for a tournament, we get that there is only one pair, which therefore is bound to play. If we had four players, then we arrive at $\frac{2}{4} = \frac{1}{2}$, indicating that half the pairs would meet, which is the case as there are *three* matches in such a tournament that starts at the semi-final stage, and there are *six* possible pairings of four players A, B, C, and D: A v B, A v C, A v D, B v C, B v D, and C v D.

This answer is correct, although the analogy with the raffle is not a complete one. In the raffle each ticket is equally likely to be a winner. In a tennis tournament this is not true of every player, and this affects the likelihood of particular pairs meeting— stronger players generally play more matches and so pairs featuring stronger players are more likely to play than weaker pairings. There is also the added complication of seedings in some tournaments, whereby it is ensured in advance that the higher seeded players will not play one another, except perhaps in the later stages of the tournament. This however does *not* invalidate the preceding analysis, as the pair in question was picked at random from a hat. It is similar to a horse race of, let us say, twenty starters. Some horses are more likely to win than others, but if you choose your horse at random, you will still pick the winner, on average, one time in twenty. The varying strengths of the horses in the field will not affect this, as you have chosen to ignore all considerations of form.

When it comes to reckoning with chance, we really have to be careful—there is no form of calculation where an answer can be so wrong without people realising it. Probability theory has always been plagued by seductive arguments that have the capacity for holding some of us totally convinced of a false conclusion. Really treacherous mistakes are often made, not only by reckless gamblers, but by sober minded judges in courts of law. Blunders often arise in quite simple situations involving conditional prob-abilites: probabilities of one event happening given that another has. A recent court case in Britain involved multiple cot deaths of infants (and similar examples have arisen in a number of countries). A woman was suspected of murdering her own children because two had died in this way. An 'expert' had said that the probability of such an event was millions to one against. Such flawed calculations often assume that each death is independent of the other, like a pair of coin tosses. This is of course not the case. This syndrome may only strike one baby in 10,000, let us say but, *given that a family has suffered one cot death,* the probability of having to endure another is much, much higher. Indeed we can expect about one family each year in Britain to suffer such a double blow.

No medical or forensic opinion would be accepted in a court of law unless it came from a properly qualified person. In the same way, no precise probability statement along these lines should be admissable as evidence unless is has been properly vetted by, let us say, an accredited representative of a recognized Statistical Society. The consequences of failing to face up to this continues to ruin lives.

On a lighter note, the same point (confusion surrounding conditional probabilities) is just as tellingly made by the joke of the man caught smuggling a bomb onto a plane, his defence being that it was to ensure everyone's safety because 'the chances of two bombs on the same flight are one in a billion!'

Some Collectable Problems on Chance

Probablility questions have their own peculiar charm for there is surprising variety in many real world problems and the paths to their solutions, although often elementary, involve sharp and novel observation concerning ratios and relationships of events. The variety of tricks on offer make the subject a kind of art where connoisseurs delight in discovering a new idea. Here follows a selection.

Invincible Teams

Consider a Knock-out Cup competition in tennis or football. The organisers set things up so that the number of teams is a power, 2^n, of 2 so that there is always an even number of players or teams at the beginning of any round. In soccer for instance the Cup Champions will be the side that survive the n rounds undefeated. We now investigate the progress of a pair, let us call them Celtic and Rangers, of otherwise invincible teams. That is to say, neither of these teams can ever be beaten except perhaps by the other. (This is none too far from the current situation that prevails in Scotland.)

Since the pairs of teams that are to play in each round of the competition are drawn from a hat, there is no guarantee of a Rangers–Celtic final. The two teams will certainly meet as no other team can stop them, but it may not be in the final round—it could be in the first or some earlier round. Nonetheless, Rangers–Celtic finals are amazingly common. It turns out that the probability that the two Invincibles of the 'Old Firm' clash in the final is better than 50–50.

Why should that be so? Let us look at the simplest cases first. Let $p = p_n$ be the probability of a Rangers–Celtic final, given that there are n rounds in the entire competition. Clearly if $n = 1$, then $p_1 = 1$, as there are only two teams to play. Next suppose that $n = 2$, so we begin with four teams, let us call them $\{A, B, C, R\}$, where C and R stand for Celtic and Rangers. There are three possible draws for the first round (not regarding which teams are home and which away), which are:

$$\{AvB, CvR\}, \{AvC, BvR\}, \{AvR, BvC\}.$$

These three possibilities are all equally likely, so that the probability that Celtic and Rangers do *not* meet in the first round and so go on to meet each other in the final is $\frac{2}{3}$. Another way of seeing this is to focus on one of the Invincibles, R say. Rangers are equally likely to draw any of the three teams, so they have a $\frac{2}{3}$ probability of missing Celtic in the first round. Therefore $p_2 = \frac{2}{3}$.

Suppose now that there are $2^3 = 8$ sides in the competition. We end with a Celtic–Rangers final if and only if they do not meet in either of the first two rounds. The chances that they miss one another in the first round is $\frac{6}{7}$ (from R's viewpoint, there are 7 other teams, 6 of which are not C). Given that they miss one another in the first round, the probability that they miss each other in the second round also is $\frac{2}{3}$, as there are four teams remaining at this stage, and we have already solved that problem. The value of p_3 is then:

$$p_3 = \frac{6}{7}p_2 = \frac{6}{7} \times \frac{2}{3} = \frac{4}{7}.$$

By extending this argument* we find that in an n-round contest the probability of a Celtic–Rangers Final is

$$p_n = \frac{2^{n-1}}{2^n - 1}.$$

The first few values of p_n are:

$$1, \frac{2}{3}, \frac{4}{7}, \frac{8}{15}, \frac{16}{31}, \frac{32}{63}, \cdots$$

We see that even for a five-round contest, the probabilty of a final between the Invincibles is nearly $\frac{1}{2}$. The limiting value is indeed $\frac{1}{2}$, although for any value of n, the probability p_n is always slightly greater than this. We really can look forward to many Rangers–Celtic finals in years to come.

Without describing all the techniques required for their solution, we continue with some additional examples of interesting questions.

Bertrand-Whitworth Ballot Problem

Two candidates in an election poll p and q votes respectively, with p greater than q. What is the probability that the winner leads the count all the way through?

The answer is delightfully simple: $\frac{p-q}{p+q}$. For example if p and q were respectively 60 and 40 votes, the probability ratio is $\frac{20}{100} = 0.2$. In other words, even though one candidate won by a handsome margin, there is a 80% probability that, at some stage at least, he was not leading his rival in the count.

I mention the problem as it is worthy of attention for two additional reasons. First, although phrased in terms of the counting of votes, being very natural and simple, it arises in many other contexts, including particle physics. Second, it is notable for its method of solution. The necessary counting is done by considering the graph of the lead of the winner throughout the count. In order to count the number of favourable paths, a particular geometric symmetry is invoked, known as the Reflection Principle. This allows us to show that the set of paths we are trying to count is equal in number to another set of paths which, although nothing to do with the original problem, are simple to count and so affords us a solution. This ingenious trick is very valuable and cannot be dispensed with in problems of this nature. What is more, these problems come in what are known as dual pairs. The answer is in fact also the answer to another related problem that arises by considering all the reversed counts of the ones of interest. In this case, the duality principle tells us the given ratio is also the probability that the winner's final margin of victory is never attained until the very last vote. This is because the reversal of a count in which the winner always leads is characterized by this latter property.

The Birthday Problem

How many people do you need in the room to ensure at least a 50-50 chance that two or more share the same birthday?

Unlike the previous problem, this is not tricky to solve, but requires quite a bit of arithmetic, so that the answer is not easy to guess in advance.* It turns out that 23 people is enough to ensure a probability of more than one half of a birthday coincidence. In practice, it is a little less, as birthdays are not quite uniformly spread throughout the year, which increases the chances of sharing. The relatively low number needed before coincidences start to pop up is the reason why most classes of children enjoy shared birthdays somewhere in the year, one of life's more pleasant little surprises.

Russian Roulette Problem

Many games are a little unbalanced in that the players takes turns and so the one who goes first has an advantage. In noughts and crosses for instance, if you go first you should never lose. It has even been proved that in the vertical connections game of *Connect*

Four the leading player has a forced win, meaning that a computer can be programmed to play the game so that it will always win provided it gets to place the first counter. This is a surprising result as the game is complicated and most skilled human players cannot achieve that level. Even in chess, playing white is a considerable advantage and Grand Masters certainly make the most of 'having the move'. In practice, at top level, a player who beats a good opponent while playing black will have won a real triumph. The best players make the tiny advantage of going first with the white pieces count for a lot and will keep their opponent on the defensive well into the game because of it. Indeed it has never been proved that white does not have a forced win in chess—it is conceivable that, with best play, the white king should always be the victor. However, that seems highly unlikely and the received wisdom is that black should be able to hold white to a draw no matter how well white plays. Careful game analysis usually reveals at least one demonstrably bad move by the loser somewhere in the course of the game.

Russian roulette is also an example of a biased game. A non-suicidal version of the game involves players taking it in turn to roll a die and the winner is the one who first rolls a six. What are the chances that the first player, A, beats the other player, B? Let a and b be the relative proportions of occasions that A and B respectively roll the first six, so that $a + b = 1$. Now A can win in one of two distinct ways. One time in six he will win on the first roll. However, if this does not happen, the tables are turned in that B now holds the advantage that A has just lost. This occurs five times in six, giving the relationship $b = \frac{5}{6}a$. Coupling this with the earlier observation that $b = 1 - a$ we obtain $1 - a = \frac{5}{6}a$, so that $\frac{11}{6}a = 1$, or in other words $a = \frac{6}{11}$. In terms of percentages, A will win about 54.5% of the times the game is played.

Some games try to even out the advantage of going first. In a tennis tie-break for instance going first corresponds to 'having the serve'. One player serves the first point but subsequently the players alternate in taking pairs of service points so that at the completion of each pair it is one and then the other who has had the extra service. It is to be noted that this attempt at balance is not applied in penalty shootouts in soccer where teams simply alternate and so there seems to be more pressure on the team that is forced to go second throughout. Since World Cup Finals have been settled by penalty shootouts, there is a lot at stake here and perhaps it is time that this format was re-evaluated for fairness. If we try this kind of alternation with our game of Russian Roulette, it does tend to balance things up, although not perfectly. If A starts with just one roll of the die, followed by two rolls by B, and then two by A, and so on, the chances that A rolls the first six is now $31/61 = 50.8\%$.*

Why Do Buses Come in Convoys?

This is a much more open ended problem. To make the setting more precise, let us suppose that a bus begins a particular route once every ten minutes. How long can you expect to wait for one to arrive?

Buses are leaving the depot every ten minutes and if the movements of each was identical, then a bus would arrive at any given stop along the way every ten minutes. A passenger who walks to the bus stop at her own convenience would then arrive somewhere in the course of one of these ten minute intervals. If she takes no particular notice of the timetable, and is just as likely to arrive at one time as another, then she is equally likely to arrive at any point during the ten minute interval between buses so that, on average, she can expect to wait five minutes until her bus comes along.

However, all experienced bus passengers have learnt that, although some days they are lucky, on average they have to wait longer than five minutes at the stop even though there may be six buses every hour. Understandably, the behavior of each bus is not identical, due to random fluctuations in road traffic and passenger numbers at stops. Some delays are inevitable, but if there are six buses every hour, *on average,* we might yet think that the *average* waiting time time remains the same as in a perfect world where a bus always arrives, like clockwork, at each stop every ten minutes.

No, it is true that *any* departure from uniformity forces up the average waiting time. Let us look at a couple of particular possibilities to get some feel for why this should be so.

If the bus route were *extremely* long, the effect of the random fluctuations that buses meet along the way will swamp the effect of the initial time difference between buses. There will still be on average six buses an hour but, very far down the route, (imagine the route actually takes years) there would seem to be almost no connection between one bus and the next—buses would just appear randomly with an overall average of one every ten minutes. In this situation, when a passenger walks up to a bus stop, the past arrival of buses is of no relevance. She arrives at the stop and looks at her watch. No matter what time it is, the *average* time till the next bus arrives will be *ten* minutes, so, in this random world, the average wait has gone from the ideal five, up to ten minutes. This is an example of what is known as a *Poisson Process,* named after the French mathematician Siméon Poisson (1781–1840), who first studied such totally random phenomena.[34] As far as the passenger is concerned, buses just materialise at random on the average once every ten minutes from the time she begins observing them.

At least this would be the case if the progress of one bus had no effect on any other. In practice, life is not so simple and, unfortunately, that tends to make things worse as the buses get in one another's way. London buses are notorious for catching up with one another and then clumping together. This really is not the drivers' fault—any delay leads to a build up of passengers who themselves arrive pretty randomly, so there is often a sudden little rush of people for no particular reason. This causes the first bus that encounters a heavy passenger load to slow down. The buses following behind tend to catch up, and sometimes overtake, but generally they find it difficult to separate once this happens. We then observe the annoying phenomenon of buses apparently traveling in convoy. At its worse we can see up to six big red London buses virtually stuck together. When this happens, we no longer have six buses per hour but, in effect, one big bus every hour. Should this happen frequently throughout the day, passengers really are only seeing one bus every sixty minutes, and their average waiting time after they walk up to their stop will be somewhere around half an hour instead of the ideal five minutes.

To understand why departure from uniformity will always make things worse, it is enough to look at the case where there is one bus

---

34 The famous data set often quoted in this context concerned the number of deaths due to horse kick in the Prussian army: Poisson theory shows that if these deaths were unrelated and occurred entirely at random, we would expect that two of them should happen in 34 of the 280 months for which there were records; the real figure was 32, and in general the data fits the 'random' hypothesis remarkably well—you just never know when a horse is going to lash out.

per unit time interval, which we take to be one hour. Suppose that a bus arrives at your stop on the hour, every hour, come rain or shine, and you, the passenger, wander up to the stop, paying no regard to the time of day. Your waiting time is equally likely to be any length of time from zero to sixty minutes, and so your average wait will be half an hour, the value exactly half way between these two extremes.

However, suppose that one bus arrives at the wrong time, either early or late. This disrupts *two* successive time periods, and if the passenger arrives at random during the two-hour interval, her expected waiting time will be longer than thirty minutes, for there are now *two* periods involved, one shorter than one hour, and the other longer. The passenger's average wait will be half the length of the time period that she enters into, but she is more likely to hit the longer period, simply because it *is* longer, and that introduces a bias towards a long wait.* Life really would be much less frustrating if buses, trains, and planes could run on time.

St Petersburg Paradox

A problem proposed by Daniel Bernoulli, in 1725, caused so much vexation throughout the 18th century that it came to be known as a paradox. Peter and Paul play a game in which a fair coin is tossed repeatedly until a head shows. If the first toss is heads then Peter wins 2 crowns, if the first is a tail and the second a head, then he wins 4 crowns, if they have to wait three tosses before seeing a head, then Peter wins 8 crowns, and so on. The question is, what should Peter pay to Paul for the privilege of playing the game? Basic probability tells us that Peter will win 2 crowns half of the time, 4 crowns $\frac{1}{4}$ of the time (because the probability of tail-head is $\frac{1}{2} \times \frac{1}{2}$),

8 crowns on average one time in 8, and so on. His average expected gain is therefore the sum of all these contributions, each of which is 1 crown. But an infinite sum of 1's is infinite! The conclusion is that no amount of money is enough to cover the expected losses of Paul.

This seemed a nonsense. To test the point, Comte de Buffon (1707–1788) made a practical experiment and played 2084 games. He found that on average Peter won something less than 5 crowns. How can theory and practice be reconciled?

The legitimate objection made by some commentators was that the game was strictly impossible to play. Paul could only offer the game to Peter if Paul had access to absolutely unlimited funds so that he could guarantee any payout. This is impossible and, in a way, this does resolve the paradox.

Indeed another version of the paradox is the argument that it is always possible to win in any casino game, as long as the player has some fixed and positive chance of winning, however small. The player makes his bet and if he loses, he plays again, this time betting enough to ensure that, if he does win, his winnings will outweigh his losses. He can in principle do this indefinitely—just keep piling on bigger and bigger bets so that, when the player finally does win, he is in front. This way he is bound to win in the end.

Again this is only correct if he has infinite funds to begin with. (And if he had, what is the point of playing in the first place?) If the player has a very large but finite fund, he can try to adopt this strategy and, if he begins with a small bet (small compared to the fortune the player has behind him), he will almost certainly end up in front. However the downside is that the amount he will win is very small compared to the fortune that he is risking and there is a small risk that he will blow the lot. This style of gaming is the opposite of a lottery. In a lottery the player sacrifices a small stake, that he will almost certainly lose, for a tiny chance of a big win. The rich man playing the double or nothing type of strategy is, on the other hand, taking a small risk on a huge stake, for the sake of being almost certain of a tiny win.

All the same, why was the gap between theory and Buffon's reality so large? The fact that Buffon did not see Peter's big win emerging in practice is also simple to explain and examples can be given that do not involve open ended games with limitless stakes. This kind of poor return will almost certainly be witnessed in any game where a big payout is possible, but very unlikely. A state lottery is a fine example. Let us say that a ticket costs $1, and the ticket holder has one chance in a million of winning a million dollars, but otherwise he gets nothing. (This is not far away from the reality of state lotteries.) Since one million dollars is put into the kitty by the players, which is all paid back to the lucky winner, this is a perfectly fair game, and tickets are being sold at the right price. However, if Buffon starts buying his weekly tickets, he could well play for ten thousand years and would still be more than likely never to win a penny.* Empirically the lottery game would seem to be worthless, for by then poor Buffon would have paid in over $500,000 for his tickets over ten millenia, only to be a loser every single time.

Buffon's Needle Problem

Buffon, who was more a naturalist than mathematician, is most famous however for his curious Needle Problem, the first problem in geometric probability. He asked if a needle is tossed on to the floorboards, what is the probability that it lands on a crack? The answer, which we will not work out here, is $\frac{2l}{\pi d}$, where l is the length of the needle, and d the width of each floorboard. If $l = d$, we see the answer is simply $\frac{2}{\pi}$. The emergence of π in the calculation is due to the fact that whether or not the needle lands on a line depends on two independent random events: the distance of the center of the needle from the nearest line, and the *angle* that the line of the needle makes with the parallel direction of the boards. It is this latter feature that introduces the circular aspect of the problem, which leads to an answer involving π. As a consequence, it is possible to estimate π by conducting the experiment many times, using the empirical ratio of successes to failures as an estimate for $\frac{2}{\pi}$, and from there find the approximate value of π itself.

In practice though, this problem does not give an accurate answer until many thousands of repetitions. Some variants of the problem do better. In the Buffon-Laplace Needle Problem, we drop needles on to a rectangular grid, and ask for the chances of a needle landing on a line (see Fig. 8.3). This experiment homes in on the value of π more quickly than the original.

A similar problem, that is easier to deal with, is that of asking for the probability that a coin rolled across a chessboard covers a corner. Again the answer depends on the relative size of the coin but let us suppose that the coin is smaller than the chessboard squares. The key to the problem is to observe that the coin will cover a corner exactly when the distance of the center of the coin to some corner is no more than the coin's radius.

Since the middle point of the coin is equally likely to land anywhere within a square, the proportion of favourable outcomes is the ratio of the favourable area within a square, divided by the total area of the square (Fig. 8.4). Since the areas that are close enough to a corner comprise four quarter circles of the same radius as the coin, we are left with a solution that is both pleasing and

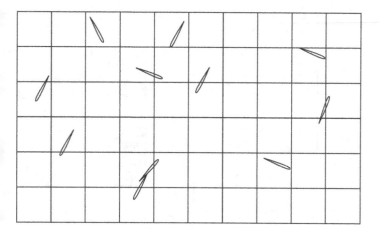

Figure 8.3. Buffon-Laplace needle problem.

simple:

$$\text{Probability of coin covering a corner} = \frac{\text{area of coin}}{\text{area of square}}$$

Bertrand's Paradox

Probability has been described as common sense put into ratios. However I have noticed that people often claim their own opinion is common sense when it seems obvious to them but not necessarily to others. By claiming that your own outlook is common sense, you implicitly charge anyone tempted to disagree with you of being a fool, even before they have opened their mouths. The record seems to show that a valid intuition on questions of chance took a long time to mature. Although simple mistakes are still made, and

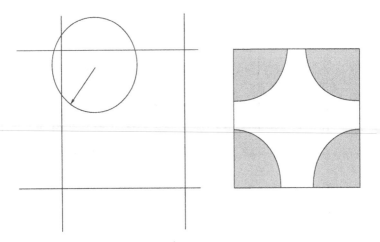

Figure 8.4. Coin on a chessboard problem.

people often harbour false expectations when it comes to matters of chance, the notion of probability is generally better understood, both by experts and the general population, than was the case in years past.

For example, the St Petersburg Paradox caused more vexation than it ever really had a right to. What is more, there was continued confusion surrounding questions of randomness to the extent that the entire field was in bad odour.

One problem that called for clarification was due to Joseph Bertrand (1822–1900). There are several equivalent versions, but in one he asked for the probability that a randomly chosen chord of a circle is no longer than the radius (which we may as well take as being one unit long). The trouble was, you get a different answer depending at how you did the choosing. One way of doing it would be to choose an angle at random, anywhere between 0° and 180° and the chord is then defined by the two points on the circumference of the circle which form the given angle θ with the center. The length of the chord will be less than the radius when the angle separating the arms is less than 60°, but not otherwise. Since the angle is assumed to be equally likely to fall anywhere in the possible range from 0° to 180°, the answer to Bertrand's question would then seem to be $\frac{60}{180} = \frac{1}{3}$.

On the other hand, we can equally well select a chord 'at random' by choosing a point along any radius of the circle and drawing the chord that lies at right angles to that radius line. A little geometry now reveals that we obtain a short chord of length less than the radius if and only if that point lies outside the shaded circle of radius $\sqrt{3}/2$ that of the original circle (see Fig. 8.5). Since the point on the radius chosen to determine the chord is equally likely to lie anywhere on the radius as any other, the proportion that leads to short chords is $1 - \sqrt{3}/2 = 0 \cdot 134$, a number which is different from $\frac{1}{3}$, and indeed considerably smaller. Which, if either, of these two approaches to the problem is right?

The difficulty in the problem was perceived to be with the nature of the infinite, however this is not really the case. It is true that we cannot consider the collection of chords of the circle the way that we would a finite collection of raffle tickets that are chosen at random, and the simple counting techniques that we use in the case of a finite problem do not immediately apply. However, the reason why we get two answers, is that we are solving two different problems. The difference in the two problems could be observed even in finite version, and that is more to do with geometry than with counting the infinite.

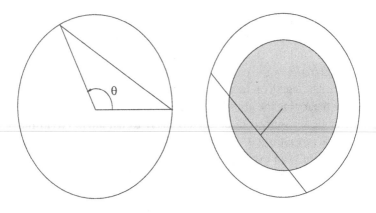

Figure 8.5. Two solutions of Bertrand's problem.

If we divide the circumference of the circle up into a large number of equal arcs, numbered from 1 to n, and choose two numbers at random from this range, we could then form a chord joining the midpoints of the arcs. For large values of n, the chord would be shorter than the radius approximately $\frac{1}{3}$ of the time as this method of selection corresponds very closely to the random selection of the angle that the chord subtends at the center of the circle. If, on the other hand, we followed the alternative approach, we would divide the radius into short uniform lengths and choose one at random to form our chord. That chord would be short approximately 13.4% of the time. I leave to the reader to contemplate which, if either of these two alternatives, corresponds most closely to the physical experiment where long thin needles are dropped at random on to a unit circle, and the length of the resulting chord is measured.

chapter 9

The Complex History of the Imaginary

The free use of algebra is the hallmark of modern mathematics. Everyone recognizes the appearance of x's and y's as the point where real math begins and mere arithmetic is left behind. Most school mathematics problems, even those of geometry, are generally reduced to equations, and their solution involves manipulation of algebraic symbols in accord with the Laws of Algebra, which are the ordinary rules of arithmetic applied to symbols instead of particular numbers. Heavy use of co-ordinate systems to tackle problems of space emphasises the drive to reduce everything to equations, and then to numbers, as quickly as possible. Even Pythagoras's Theorem, the sum of the squares on the shorter sides of a right triangle is equal to the square on the hypotenuse, is usually summarised as the equation: $a^2 + b^2 = c^2$, a form of thinking that would have been altogether foreign to the ancient Greeks.[35]

The triumph of algebra, which may be rather too overwhelming for the good of the subject, is nonetheless understandable. For mathematics to be applicable in any sense at all we need to be able to do something with it. In practice this nearly always means developing forms of calculation, and this imperative channels its practitioners into algebraic manipulations of one form or another and ultimately into producing numbers. To the modern mind, this might seem natural and inevitable. What might be difficult to appreciate therefore, is why the rise of algebra proved to be so slow and hesistant.

Part the reason undoubtedly is that people are afraid of mathematics, and here I am not only referring to those of us who find it very difficult. This fear applies in much the same way to gifted mathematicians as it does to the mind of the mathematically ordinary person. *Everyone* has their own way of thinking about the subject: their own intuition if you like, and everyone's outlook is limited. Once mathematics takes a strange turn, the first impulse often is to abandon the road, and return to a safer path, more in company with our own way of thinking. There are numerous examples of first rate mathematicians turning their back on a good mathematical direction because they did not like the look of it.

Centuries of struggle have taught us the importance of being open-minded and that certainly has informed the modern outlook.

---

35 The first to mix algebra and geometry in this fashion in order to exploit the methods of the Arabic algebraists was Regiomontanus (1436–1476) although he expressed his methods in a purely rhetorical fashion without the benefit of algebraic symbolism. His work languished in obscurity until a century after his death, but was very influential from around 1575, when the Latin version of his *Arithmetica* was published.

However, it is not always clear what will or will not be a fruitful area of research, and this is where subjectivity enters into the subject. For example, the program to find all the so-called sporadic simple groups mentioned in Chapter 1, was a colossal and sustained effort, involving many mathematicians across the world over a number of years. Is the problem worth that much energy, or should these people be turning their unquestioned talent in directions likely to yield more? Only time will tell, but my own view is that we should allow the best people to pursue the questions that capture their imagination and trust to their judgement.

Established figures in a field often give well received lectures as to future directions of the subject. These can serve to galvanise research in a coherent and productive way. The most outstanding example in mathematics is that of David Hilbert who, at the Paris Mathematical Congress in 1900, set a list of 23 problems to which the mathematical community was exhorted to turn its attention. Another earlier example along these lines was the Felix Klein *Erlanger Program*, which pointed the way for the thorough going study of invariants in geometry.

Some more recent attempts at directing the mathematical traffic have proved less enduring however, as the opinions of even the most eminent individuals can date surprisingly quickly nowadays. In addition, directions that look likely to be rewarding, can lead to deep frustration. An example is the way in which patterns that evolve from very simple rules can reveal incredible structure. Computers and modern visualisation techniques have done much to draw researchers into these lines of investigation. Seeing is believing, and fantastic pictures can hold our attention as little else can. However, adequate mathematical description of the phenomena observed can prove desperately elusive. No-one wants to waste years chasing rainbows, but at the same time everyone is reluctant to turn their backs on a truly interesting topic when it seems that we must all be missing the point.

Algebra and its History

How does all this square with the development of algebra? First let us say a little about the nature and history of the subject.

Algebra originated from problems involving a search for a number, or numbers, of unknown value.[36] Most of us meet up with problems of this kind somewhere in our school career. Gilbert and Sullivan's model of a modern major general sang the boast that he was 'very well acquainted with matters mathematical and understood equations both simple and quadratical'. By a simple equation the general meant a *linear equation,* that is one whose graph is a straight line. An example of this is the conversion formula from degrees Celcius to Farenheit: $F = 1 \cdot 8C + 32$. Problems involving this kind of linear relationship where one quantity is related to another by means of a scaling factor ($1 \cdot 8$ in this case) and a shift (of 32 here), are indeed pretty simple. For example if we want to find the temperature where the two scales agree, we simply put $C = F$ in our conversion formula to give $C = 1 \cdot 8C + 32$, which tells us that $0 \cdot 8C = -32$, so that $C = -40$:

---

36 An astonishing insight from the Persian poet and mathematician Omar Khayyam (ca. 1100 AD) cannot be allowed to pass, "Whoever thinks algebra is a trick in obtaining unknowns has thought in vain. No attention should be paid to the difference in appearance of algebra and geometry: algebras are geometric facts which are proved."

the temperature of $-40°$ is the unique value where the two scales agree.

All the same, even linear relationships can cause real confusion among people who should know better. For example, it is important to appreciate the distinction between saying that a *difference* of $100°C$ equals a *difference* of 180 degrees Farenheit (which is true) and saying that $100°C$ is the same temperature as $180°F$ (which is false—the corresponding Farenheit reading is $212°$, as the conversion formula readily reveals). This exact point went missing in an article on global warming in a well respected daily newspaper. We were told to expect an increase in the ambient temperature in the coming years of $0 \cdot 4°$ C, and readers who preferred to think in Farenheit were assured that this was equal to $32 \cdot 7°$ F! The given Celsius increase is too small to be noticed on a day-to-day basis, while a general rise of $32 \cdot 7°$ F would melt the polar ice caps and destroy most life on the planet! To avoid such blunders, newspapers would do well to employ some people who understood a little mathematics, for such a basic mistake in a science article is a severe embarassment.

To put the record straight, an increase of $0 \cdot 4°$ is the same as an increase of $0 \cdot 4 \times 1 \cdot 8 = 0 \cdot 72°F$. What the writer of the article had done was to treat the $0 \cdot 4$ as a *reading* on the Celcius scale, as opposed to an increase, and then converted that reading to the Farenheit Scale using the above formula in order to come up with the figure of $32 \cdot 7$.

Even more simple mistakes can be found however: recently I read of a newly discovered planet, photographed by the orbiting Hubble telescope, and was assured that it was 3,100 trillion light years from Earth. It seems the editor believed that no distance was too great for astronomy, but it is so easy to write down huge numbers without giving a moment's thought to whether or not they make sense. The entire universe is not that big—there is nothing that could meaningfully be described as being 3,100 trillion light years from the Solar System.

The explanation for the error is simple enough. The star system in question is about 500 light years distant (very much within our own sector of the galaxy) and this figure had been converted to miles, while forgetting to change the units involved.

Perhaps I shouldn't be so judgemental, as there are many questions involving only simple linear relationships between one quantity and another that are very testing all the same. My own favourite is *Sam Loyd's Ferry Boat Problem.* This comes from the collection of mathematical puzzles of the 19th century American writer, Sam Loyd.

Two ferries, moving at constant but different speeds, start out simultaneously but from opposite sides of the river, and first pass each other 720 yards from one bank. Each has a ten minute change over period in which passengers disembark and the new ones alight for the return crossing. On return, they once again pass each other, this time from a point 400 yards from the other shore. The question is, how wide is the river?

One might begin hopefully by writing 'let w denote the width of the river'. We want to find the number w, but the information given is peculiar. The extra complication about the change over does nothing to clarify matters. One piece of practical advice that can be given about a problem that seems too complex to tackle at first sight, is to try first to solve the simpler question where the complication is ignored. If we can do this, we have made real progress, and can reasonably hope to return to deal with the full problem later. Having said this, let us ignore the change over period or, more precisely, let us assume for the moment that it is zero, so that each boat reaches the opposite bank and then bounces off, traveling the return journey at the same speed as before.

When the boats first meet, one of them has traveled 720 yards. When they next pass, the same boat has traveled one full width plus 400 yards: that is a distance of $w + 400$. How are these two distances related?

The aspect on which to focus is the *total* distance traveled by the river boats. When they first meet that total is w, for together they have traveled the width of the river. When they meet again, each has completed a single crossing, contributing $2w$ to the overall total, and their combined part return trips represent another river width w, so that the sum of the distances sailed is now $3w$. Since the speed of each boat is constant, this observation applies to each, so that *when they cross a second time, each has traveled three times the distance they had traveled when they crossed the first time.* This answers the question posed at the end of the previous paragraph: $w + 400 = 3 \times 720 = 2160$ and so $w = 2160 - 400 = 1760$ yards. We discover that the width of Loyd's river is exactly one mile.

Ah, but what of the ten minute break that we conveniently forgot about? If you look at our explanation, you will see that the change over complication was merely a distraction, as the key observation in italics still holds, even allowing for the common change over period—when they next meet, each ferry will have traveled three times the distance it had after their first passing. This conclusion continues to hold, no matter what the length of the transition, provided only that it is the same for each ferry.

The equation that we elicited for the unknown quantity w was very simple indeed, however the argument that justified it was quite subtle all the same.

Right from it earliest beginnings in Mesopotamia around 4,000 years ago, we see that the Babylonians delighted in recording their clever methods for solving problems like Sam Loyd's riddles that could have had limited practical value. In a tradition that has lasted up till the present day, the puzzle maker would place the reader in a familiar setting, typically involving a market transaction or establishing the extent of a field, and then call for the answer to a question that was often artificial in the extreme. Of course it was neither the question, nor the answer that was the important issue, but discovering the way to go about solving it. Mathematics often forges way ahead of its applications. Placing a mathematical principle for the purposes of instruction in an easily appreciated setting is a little art in itself.

The high point of sophistication of basic algebra is the quadratic equation: an equation involving an unknown value x that appears in squared form, x^2, in the problem to hand. The Mesopotamians enjoyed posing such problems but only positive solutions were recognized as meaningful.[37] Typically they might call for the dimensions of a rectangle given that its area is say, 36 square units and its perimeter is 30 units. Our approach to this problem would be to label the sides of the rectangle by its yet unknown lengths, x and y, and write down the given information in the form of a pair of equations. Substituting from one into the other leads to a quadratic equation in x that we would solve to find we had a

---

37 The first instance of an isolated negative number in an equation occurs in *Triparty en la science des nombres* by Nicolas Chuquet around 1500: .4^{1}egaulx a $\overline{m.2.^0}$—that is $4x = -2$.

12×3 rectangle.[38] The Mesopotamian scribes lacked any kind of algebraic notation, so their solutions always referred to the particular problem at hand, and the method of solution was always given as a recipe applied to the problem, which read like the list of instructions given to the listener in our 'guess the number' tricks of Chapter 3. 'Do this, and it will work' the apprentice is told. However, they had no clear way of explaining what 'this' was. One example then would not be enough to make the general method clear, and so they would pose, and solve, a long list of similar problems, until the reader would get the hang of how to do them.

It is easy to imagine a student asking in frustration for an explanation of where the steps were coming from, as the description seems to pull the required numbers out of thin air. In practice, the learner would have needed a teacher on hand to coax him through. The underlying approach though was sound—the Babylonians solved their quadratic equations through the method we call *completing the square*. The name arises from the geometric interpretation that we find in Euclid's *Elements*, and is the basis of the modern quadratic formula, which is the most complicated piece of algebra that school pupils are expected to memorize.

A problem that comes down to solving a linear equation might be solved by an intelligent person with no mathematical training just using their raw mathematical wits. They would need to be clever, and they might have trouble explaining how they arrived at their answer, but they could solve the problem all the same. By contrast, there is no other way to solve a general quadratic equation, except through somehow making use of the technique

---

38 The equations are $x + y = 15$ and $xy = 36$; substituting $15 - x$ for y in the second of them then yields $x^2 - 15x + 36 = (x - 3)(x - 12) = 0$, which gives, $x = 3$, $y = 12$ or, what amounts to the same rectangle, $x = 12$, $y = 3$.

of completing the square, and this is not something you could reasonably expect anyone to discover by themselves. It is the first piece of genuinely difficult algebra. The well known quadratic formula has the method already built-in: although the user can get the answers without knowing about the algebraic technique on which the formula relies, it is still there in the background working on his behalf. This modesty leads to mathematics, and mathematicians, not always getting the credit they deserve. A great deal of mathematics is embedded in the software of everyday life, and works invisibly behind the scenes for the benefit of everyone.

Solution of the Cubic

Throughout the first half of the 16th century, tentative steps towards a modern arithmetic and algebra are to be seen in works of various French, German, and Italian mathematicians. For instance, the German Michael Stifel (1487–1567), at the turn of the 16th century, showed a sound grasp of the arithmetic of negative integers, although he still referred to them as "numeri absurdi" as he appreciated that although they could be useful in formal manipulations, it was unacceptable to think of them as existing in their own right. Despite these encouraging signs, the most difficult algebraic problem dealt with was still the quadratic equation, which had been understood for the best part of 4,000 years. Certainly cubic and quartic equations, which involve unknowns raised to the third and fourth powers respectively, were untouchable. Indeed, in the early 12th century, Omar Khayyam had expressed the view that cubic equations were algebraically insoluable, so their solutions could only be represented geometrically through points of inter-section of related curves.

This settled pessimistic view was overturned in a moment when, in 1545, the solution of both general cubic and quartic equations was published in the famous book of Geronimo Cardano entitled *Ars magna* (The Rules of Algebra). This sudden and unexpected leap in understanding was of such psychological magnitude that 1545 often is taken to mark the beginning of the modern era: the discovery of the New World of Mathematics.

Cardano was not the one who discovered the techniques, as he freely admits in his book. The substitution devices that allowed any equation of the fourth degree to be reduced to a cubic, were discovered by Ludovico Ferrari (1522–1565) who was Cardano's loyal student.[39] The first to discover a method that applied at least to some types of cubic equations was Scipione del Ferro (ca. 1465–1526), a professor of mathematics at Bologna, the oldest university in Europe, and about whom little else is known except that he bequeathed the secret of his technique to a trusted student of his own, Antonio Maria Fior.[40]

The two principal protagonists in the claim to priority however are Cardano himself and Niccolò Tartaglia (1500–1557). After a long correspondence, Tartaglia (1500–1557) reluctantly divulged to Cardano his method for solving any cubic equation. This was done in teasing verse, whilst at the same time he exacted a solemn oath from his correspondent never to publish the secret technique.

39 Claiming a form of paternal glory, Cardano writes that the technique 'is due to Luigi Ferrari, who invented it at my request.'

40 del Ferro's papers were rediscovered only in 1923 in the library of the University of Bologna by Ettore Bortolotti. The date of his discovery in now put at 1515, about ten years later than was previously thought.

It was normal in these times to guarantee one's reputation through demonstration of superior knowledge, while not revealing its source.[41] There was still an element of the high priest in a leading scientist or physician. Tartaglia stubbornly refused to publish the extent of his knowledge on the cubic, apparently because he felt it in his own best interests not to do so. His reluctance is still hard to understand all the same and in the end Cardano lost patience and published what he had. Tartaglia's relationship with Cardano, which was always prickly, could not recover from this betrayal.

However, it should be added that Tartaglia himself is not above reproach for he had claimed more credit than was his due in some of his earlier work on other subjects. Indeed it is possible that he gleaned an inkling of how to solve the cubic from some other source—at the very least he would have heard the rumours that circulated suggesting the problem of the cubic might be algebraically solvable after all.

Historians of mathematics have, up until fairly recently, been quick to condemn Cardano for his behavior. However, perhaps because such things as oaths are not taken as seriously as they once were, the modern appraisal of these characters is less judgemental and the attitude towards them has become ambivalent. Through their writings, Cardano comes over as a more agreeable character than Tartaglia, something that counts for more than it once did. What is certain is that Cardano did the mathematical community a favor by the publication of these new techniques. Scipione del Ferro, about whom we know least, is perhaps now the most admired for his pioneering contribution.

41 In 1535 Fior publicly challenged Tartaglia to a problem-solving contest, based on cubic equations, only to be humiliated by the superior mathematician. Many a man sets out to shear but comes home shorn!

All the personalities in this drama were complicated characters who experienced great personal difficulties. Tartaglia's real name was Fontana. The name by which he is known means stammerer, an affliction that befell him through a sabre slash he endured as a child at the fall of Brescia to the French in 1512. Ferrari was a fiery hothead who lost a few fingers in a fight and was probably murdered by his sister who was keen to inherit the little money he had earned. Cardano himself was an illegitimate child born only as a result of a failed attempt at an abortion. He eventually succeeded in becoming a sought after physician who successfully treated the Archbishop of St Andrews in Scotland. (Unfortunately, his patient lived only later to be hanged.) He wrote hundreds of books on all manner of subject and some made him enemies such as *On the bad practices of medicine in common use.* He lived in a house surrounded by animals, of which he was fond, and his impossible children.

What the reader may well be imagining is that Cardano had published a formula, like the quadratic formula only more complicated, for solving equations involving cubic terms. However this is not an accurate description. First, there is the basic point that modern algebraic notation had yet to be developed, and so the method had to be described largely in words.[42] Even the basic device of representing the unknown by a single symbol, x or a or whatever we may choose, was not in use. Indeed there was not even an adequate word for the object of discussion—the unknown constantly being referred to as the *cosa*, which is simply the Italian word for

---

42 The word *algebra* derives from the book *Al-jabr wa'l muqabalah,* written by al-Khāwarizmī, from whose name comes our word *algorithm* for describing a mechanical list of instructions. Despite this, there was no algebraic notation in the classical arabic works.

thing. For this reason algebra was sometimes called the *Cossick Art*. Everything was described rhetorically, so that Tartaglia's poem on the cubic that he wrote for the benefit of Cardano would not have been regarded as especially eccentric.

It was not only the language however that made the matter so long winded. By making free use of negative numbers, we can express any cubic equation in a single standard form and we could, if we wished, write this general expression out in full in ordinary language. In the 16th century however it was taken as read that the coefficients of the unknown (the multipliers of the 'thing') had to be positive, and so a raft of cases arose depending on whether certain terms appear on the left or right of the equation. Each case would have to be illustrated by a particular example, after the age old fashion. That example however was taken to be fully representative of its case. For instance, Cardano wrote 'Let the cube and six times the side be equal to 20': in modern notation he was asking us to think about the equation $x^3 + 6x = 20$. Cardano's recipe for solution was then meant to be a prototype for any equation of this form where 'cube and thing is equal to a number'. However, the equation $x^3 = 6x + 20$, would be thought of as having a different nature and the design of a solution for this type would appear elsewhere.

Cardano did not reject negative numbers completely, but still saw them as fictions all the same. It was understood that only positive solutions were of any interest, and the negatives that arose in the course of calculations were referred to as 'numeri ficti'. Even in modern mathematics, it is perfectly legitimate to focus on solutions of an equation of a prescribed type. When doing so we sometimes say the topic is *diophantine*, after Diophantus who, around the 3rd century AD, introduced clever ways of finding one rational

solution to a problem from another. Furthermore, in many mathematical applications, only positive solutions have a meaning, and
so the context of the problem narrows the feasible field of solution,
even though we may, for mathematical convenience, calculate in a
wider arena, only to discard the 'extraneous' solutions that might
thereby emerge.

This however was not all as, to 16th century mathematicians,
the many cases of the cubic seemed to represent as many genuinely
different scenarios, some of which were still very vexing indeed. To
understand this, it is worth taking a moment to reflect on how even
quadratic equations show different patterns of solution.

An equation such as $x^2 - 3x + 2 = 0$, has two solutions, namely
the integers 1 and 2, as is readily checked. Any quadratic equation
has at most two solutions as is seen, for instance, through the
famous quadratic formula.* A way of visualising this is by plotting
the graph of $y = x^2 - 3x + 2$, (see Fig. 9.1(a) below). The solutions of the above equation correspond to the places where the
curve crosses the $x-$axis, as they represent the values where $y = 0$,
and so the equation is solved. Some quadratic equations, such as
$x^2 - 2x + 1 = 0$, have but one solution, in this case the number 1
(see Fig. 9.1 (b)). This is manifested in the corresponding graph
just touching the x-axis at one point. Other quadratic equations,
such as $x^2 + 1 = 0$, have no real solution, nor does there seem to
be a need for one, as the corresponding graph sits above the x-axis
(see Fig. 9.1 (c)).

How does this different behavior manifest itself in the quadratic
formula? In the first case, the formula gives us the two solutions
in the form: $x = (3 \pm \sqrt{1})/2$: we get the solution 2 when we take
the plus sign, and 1 when we take the minus. For the second case,
it gives us $x = (2 \pm \sqrt{0})/2$, and whether we take the plus or the

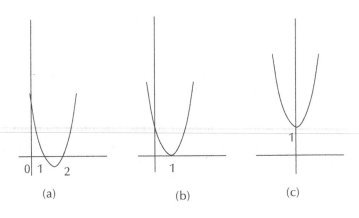

Figure 9.1. Graphs of three quadratic
curves.

minus sign, the answer is the same, the unique solution is the
number 1. However, if we peversely insist on using the formula
in the third case where there manifestly is no solution, it tells
us that the answer is $x = \pm\sqrt{-1}$. Since there is no square root
of minus one, this is a slightly disturbing development. It seems
safe to ignore it however, for the formula does work in that it
always gives all the real solutions to a quadratic equation, including
negative ones. It seems that if we apply it outside its proper realm, it
generates solutions without meaning. Since there are no solutions
in these circumstances, this is hardly surprising. It is never the less
untidy that the mathematics is suggesting there is an answer, when
we feel sure that there is none. However, there is nothing forcing
us to look any further. The purpose of the formula is to find all
solutions to a given quadratic equation, and that is a job it will
certainly do.

However, as Cardano and his contemporaries found, when it comes to cubic equations, square roots of negatives arise in a much more insistent fashion, which is difficult to dismiss. In the case of the quadratic, the standard technique only starts outputting i, which is the symbol used to denote $\sqrt{-1}$, when there was no other real solution, positive, negative, or zero. However, Cardano found that his method would sometimes give rise to a solution being expressed in terms of i, *even when he knew the solution to be an ordinary positive integer.* For example, the equation that we would write as $x^3 = 15x + 4$ has only one positive solution, which is the number 4. Cardano realized that an equation of this kind, where all terms on one side of the equation were higher powers than those on the other, only had one positive solution. The expression his method provided for the unknown value however had the square root of negatives inextricably involved. More generally, he recognized the *irreducible case,* where there were three real solutions of the given equation, as being problematic, for his method only furnished a complicated expression involving square roots of negatives. Indeed the form provided is useless for most purposes.[*43]

It is extraordinary how so momentous a mathematical discovery could in so many ways be a false dawn. It was important, but not for the reasons you might expect. The method certainly was not to be used in any practical calculation. Indeed, a century before, al-Kashi of Samarkand developed methods that could solve

---

43 Rafael Bombelli (ca. 1526–1573) bravely took on the Cardano expressions, and making use of what we would call conjugates of complex numbers he showed that Cardano's formal expressions were consistent with the known root of the equation. In order to carry out his manipulations however, he needed to know what the root was in advance: any attempt to find the root algebraically using his approach led you back to the same type of equation with which you began!

cubic equations quite adequately.[44] Cubic equations do indeed arise in practice, and are solved by numerical techniques, like those of Horner's method, whereby better and better approximation to the solutions are found by calculating and re-calculating the answer, starting from a well chosen initial guess. For example Barker's equation governs time and angle in a moonshot, and is a cubic equation in t, the tangent of half the angle determining the direction of flight. This is a very delicate equation, as it involves small differences of large numbers. For that reason special mathematical tricks had to be devised for the Apollo program so that the primitive computers of 1969 did not get it wrong due to accumulation of rounding error.

What is more, as we have seen, if the equation has an integer solution, the Cardano method may present the answer in a way that hides this. However, in the case where the equation does have at least one rational solution, there are very simple techniques for finding all the roots of the equation.* It is these that are normally taught to modern mathematics students rather than the Cardano approach.

Of course, it was the theoretical advance that captured everyone's imagination. Even here however, the new direction that seemed to be indicated turned out to be a dead end. Ferrari had shown how quartic equations could be reduced to cubic equations through his clever substitutions. It was natural to expect that an extension of his technique would allow the *quintic* to be solved: that is the equation involving powers as high as five. Surely a similar trick could be found to reduce such a problem to a fourth degree

44 The iterative method of al-Kashi, now known as Horner's Method, probably came to Samarkand from China where *fan fa,* as the technique was named, was invented by the 13th century mathematician Chu Shih-chieh.

one that could be reduced further, and solved. From this point we might expect that a general pattern of substitution would be described that would reduce any polynomial equation to one of lesser degree, whereby it would be possible, in principle, to solve any equation of the nth degree, in terms of algebraic expressions involving roots up to nth. This natural expectation endured for centuries, but turned out to be entirely false: in general, equations of degree higher that four cannot be solved in this way, a fact that was not proved until the early 19th century.[45] Perhaps the greatest legacy of the solution of the cubic was the arrival, without invitation, of the *imaginary number i* into the world of mathematics.

---

45 By the Norwegian genius, Neils Abel (1802–1829) at the age of 19. His method however was superseded by one who died even younger, Evariste Galois (1811– 1832). Galois theory has become one of the cornerstones of modern algebra. The first serious attempt at proving the insolvability of the quintic was published in 1799 by Paolo Ruffini (1765–1822).

From Imaginary to Complex

The latter part of the 16th century marked the rapid and undeniable rise of Mathematics. From around this time, the development of the modern subject shifted into full swing, with scientists
expecting, and securing real progress on many fronts, a process
that has continued to the present day. By the late 1500's we see,
in addition to the full blooded use of decimals, the advent of logarithms by the Scot John Napier (1550–1617). This was a major
practical tool of the sciences up until 25 years ago. Logarithms
exploited the Laws of Indices to turn complicated multiplication
and division sums into additions and subtractions, and the slide
rule was their physical manifestation. Although they now seem
so quaint and out-dated, logarithm tables helped 17th century
astronomers track the orbit of the Moon and, in 1969, allowed a
man to walk upon it.

Napier was inspired to pursue his idea when he heard of
the calculations of the Danish observational astronomer, Tycho

Brahe (1546–1601). The Danish School (which was forced by circumstances to re-locate in Prague) made use of the method of *prosthaphaeresis* to turn multiplications into additions through use of trigonometric equations that converted products into sums. These ideas date back to the medieval arabic astronomers of the 11th century, but had evidently taken the best part of five hundred years to penetrate into Europe.[46]

With the rise of complex arithmetic came the advent of algebra. Much of the credit here goes to the Frenchman, François Viète (1540–1603), who emancipated algebra from the geometric style by introducing algebraic unknowns that were manipulated in accord with the rules that governed arithmetic in something like the modern fashion. Viète's algebra still fell some way short of modern notation however: for instance he would write A^3 as A cubus.

Algebra rapidly became a tool of general mathematical discourse, for much could be revealed without constant recourse to diagrams and physical interpretations of all quantities in terms of areas and volumes. What is more, passing from prose to symbolic mathematics brought with it more significant benefits than mere brevity. Algebraic symbols carry a universality of interpretation that allows them to be manipulated in a way that words cannot. Indeed, this was the key breakthrough that allowed mathematics to flourish in a way that was not possible until the advent of

46 Napier's Logarithms were not the base 10 logs that eventually became standard but were closer in nature to the natural logarithms that involve the number e. The idea of base 10 logarithms was formed in conjunction with Henry Briggs of Oxford who compiled the first table of common logarithms, as they came to be called, in 1617. Another simple technique for turning multiplication into addition is the Quarter Square Rule.*

algebra. All higher mathematics relies on constant use of algebraic manipulation and would be impossible without it.

The use of co-ordinates in geometry dates from around the 1630's with the work of Fermat and René Descartes (1596–1650), although it is the efforts of the latter that came to be widely known. This began the tradition of turning geometric problems into algebraic calculations, which is the way most school and undergraduate mathematics is conducted today.

The geometrical interpretation of mathematics is a revered western tradition dating back to Pythagoras, Euclid, and the Ancient Greeks, which is still respected and persists in modern thinking. European mathematics often seeks the visual, and has been geometric in motivation and style. It has often been observed that Asian, and in particular Indian mathematics, was never so hidebound and at times showed a much more algebraic bent. It is ironic that the extraordinary high quality of classical Greek mathematics may have acted as an inhibiting factor in the development of European arithmetic and algebra. This cultural bias still persists: western mathematicians like to be able to *see* their mathematics, and algebraic triumphs over geometry are not always applauded. In contrast, perhaps the greatest Indian mathematician of the 20th century was Srivinvasa Ramanujan (1887–1920) who was very much in the Indian manipulative tradition of Brahmagupta (7th century) and Bhāskara (12th century). Mathematicians still marvel at Ramanujan's genius, and have difficulty imagining how he approached his mathematics. Whatever his thought patterns were, they were not geometric.

As early as 1500 Chuquet had run up against imaginary solutions but dismissed them as fictions that had no place: 'Tel nombre est ineperible'. Cardano on the other hand, was genuinely disturbed but could see no way forward. The imaginary roots were for him 'subtle and useless'. Bombelli had some measure of success in manipulating them but characterized his own efforts as merely 'a wild thought'.

The next notable episode was in Amsterdam 1629, which saw the publication of Albert's Girard's book *Invention nouvelle en l'algebre.* Since ancient times it was known that simple linear equations had a unique solution, but quadratic equations generally had two numbers that satisfied them. Girard seemed to be the first to appreciate that the number of solutions of an equation equalled the degree of the equation, so that a cubic has three solutions, a quartic four, and so on. However this simple statement was only true if negative and imaginary roots were recognized. By allowing for imaginary roots, the general principles of the formation of the solutions of equations, and the way in which the solutions were connected to the coefficients of the equation, could be formulated in a staightforward way that applied in all cases. Girard was using the imaginary as a framework in which to elucidate general rules that otherwise were obscured by a mass of seemingly disparate cases.

Girard however was pointing mathematics in a direction that it was not yet ready to go and the topic of the imaginary soon drifted back into obscurity to be practically forgotten.

The latter part of the 17th century was dominated by Isaac Newton (1643–1727), who introduced mathematics into physics in a revolutionary way. He was not alone in some of his works however for Gottfried Wilhelm Leibniz (1646–1715) in continental Europe also invented the methods of differential and integral calculus, and used them to good effect. All the same, everyone still was wary of the *imaginary unit i,* the square root of minus one. Leibniz toyed with the idea again in the 1690's, performing some surprising formal calculations, including a factorization of positive numbers using imaginary factors. This raised some eyebrows, yet even Leibniz distanced himself from his own mathematical musings by comparing imaginary numbers with the Holy Ghost, in that they occupied a shadowy world between existence and non-existence.[47]

The Imaginary World Is Entered

The greatest mathematician of the 18th century was the Swiss born Leonhard Euler (1707–1783). Euler single-handedly drove every important direction in mathematics throughout his long and exceedingly productive life. His kind and generous character has endeared him to subsequent generations in a manner that, due to the nature of history, will never be repeated for it is no longer possible for a single person to command all of mathematical science the way Euler did. For these reasons, Euler is fondly remembered by mathematicians in a way that some other great figures are not.

To Euler we owe the notations for the numbers π, e, and i for the square root of minus one. Calculations involving imaginaries had begun to emerge in another setting, that of taking logarithms of negative numbers. Euler introduced formal equations involving imaginary numbers that proved genuinely useful in understanding the state of play in this field. Towards the end of the 18th century the use of imaginary numbers was quite widespread.

---

47 It is natural to condemn mysticism in mathematics as a complete waste of time or worse. However individual mathematicians claim to have made discoveries through reveries on the nature of the supernatural: Kurt Gödel's solution to Einstein's equations in general relativity and Alan Turing's trail blazing work on artificial intelligence are two examples.

Nonetheless, some hesitation persisted. After all, the very word *imaginary* betrays ambivalence, and suggests that in our heart of hearts we do not believe these numbers exist. On the other hand, by calling every number representable by a decimal expansion *real,* we are making the psychological distinction more stark. Indeed the adjective *imaginary* is a somewhat unfortunate one—although an intriguing name, some students' perceptions are so colored by the word that they consequently fail to come to grips with the idea.

In order to have a number system that contains all the ordinary real numbers and the imaginary i, we must allow for addition and multiplication of all the numbers involved, and this immediately leads to the notion of a *complex number*: one of the form $z = a + bi$, where a and b are ordinary real numbers called respectively the *real* and *imaginary* parts of the complex number z. (Note that the imaginary part, as it is called, is itself a real number, the number b.) In 1797 Caspar Wessel (1768–1818) took the natural step of representing the number z as a point in the plane, with rectangular co-ordinates (a, b). The addition and multiplication of complex numbers then become very natural operations in ordinary geometry (as explained further in the next section). It was this visualisation of the hitherto mysterious imaginary and complex numbers that led to the remaining reservations as to their use finally being set aside. Every point in the plane could be regarded as representing a complex number, and vice versa. It was around the beginning of the 19th century therefore that complex numbers were finally welcomed as fully respectable citizens of the lexicon of mathematical ideas.[48]

48 The detailed history is messier: in 1806, J.R. Argand published an account of the graphical representation of the complex numbers, and the plane, when regarded as the home of complex numbers, is often referred to as the Argand plane. However,

One very reassuring outcome of the research of Gauss and others was the realization that the set of complex numbers was complete, in a way that no other previous number system could claim. The story of numbers began with the counting numbers, but these were inadequate even for ordinary arithmetic, as the two operations of subtraction and division take us out of the system, leading to the formation of the rational numbers. That system is adequate for ordinary arithmetic but, as Pythagoras proved, we still do not have enough numbers to take square roots. What is more, limiting operations, which are the lifeblood of calculus, lead us further out of the realm of algebraic numbers into the system of the reals—the totality of all numbers that can be represented by decimal expansions.[49] This is not true of the rational numbers: the limiting value of a sequence of rational numbers may be irrational. For example, we can write down a sequence of rational numbers that get ever closer to $\sqrt{2}$: 1.4, 1.41, 1.414, 1.4142, $\cdots$; however the limiting value of the sequence is not itself a rational number.

The real number system was still inadequate nonetheless, as it was not even closed under the taking of square roots, and so needed to be expanded to the set of complex numbers in order

both Wessel's and Argand's accounts were largely ignored until the leading figures of Augustin-Louis Cauchy (1789–1857) and Karl Freiderich Gauss (1777–1855) popularised them years later. Girard had already introduced the idea of the one-dimensional number line, and the English mathematician John Wallis had suggested in the 17th century that purely imaginary numbers might be represented by a line perpendicular to the axis of the real numbers.

49 This way of looking at the real numbers naturally springs to mind, but has substantial shortcomings. A consistent formulation of the real number system was not struck until the latter part of the 19th century by J. W. Dedekind (1831–1916). The so-called *Dedekind cut* (later simplifed by Bertrand Russell) resolved the apparent conflict whereby the real line, while consisting of numbers, which are discrete entities, nevertheless forms a continuum.

to allow mathematics to follow its natural path. However, upon reaching **C**, the set of complex numbers, we finally arrive in a setting in which old patterns of inadequacy are not repeated. As with the real numbers, the operations of arithmetic can be carried out on the set of complex numbers **C**, and we remain within **C**. Moreover, the limit of a converging sequence of complex numbers is another complex number. Additionally however, the square root of a complex number is itself another complex number and, more generally, any polynomial equation of degree n, has n (complex) solutions, so there is no need to travel outside of the system in search of new numbers to represent solutions to our problems. We have all the numbers we could ever need—mathematics had finally discovered its natural setting.

What is more, there is a lot of mileage in the idea that a complex number is just a pair of real numbers: that is to say we can represent the complex number $z = a + ib$ by the ordered pair (a, b). In this way, we never need mention the peculiar new symbol i if it in any way upsets us. The ordinary real numbers are subsumed into this larger set, for the real number a is now represented by the pair $(a, 0)$. The imaginary unit i is still there of course: its coordinates are $(0, 1)$.

A second insight is that there is nothing unprecedented in this. On the contrary, the passage from the integers to the rationals involves the same kind of process, where we take numbers, and form new ones, by taking pairs: the fraction $\frac{2}{3}$ is just a particular way of considering the ordered pair of numbers $(2, 3)$.[50]

---

50 Fractional notation was used first by the Greeks, at first with the denominator on top, and later in the modern fashion, but without the separating bar. However the preference for unit fractions persisted in Europe well into the second millenium.

The arithmetic of complex numbers presents itself very nicely in the complex plane, but is not without one or two surprises. Addition is certainly simple enough. When we add two complex numbers $z = (a, b)$ and $w = (c, d)$, we simply add their first and second entries together, to give us $z + w = (a + c, b + d)$. If you are happy to use the symbol i, we can give the example $(2 + 3i) + (4 + 5i) = 6 + 8i$.

This corresponds to what is known as *vector addition* in the plane, where directed line segments are added together, top to tail. In this case, starting at the *origin*, which has coordinates of $(0, 0)$, we lay down our first arrow from there to the point $(2, 3)$. To add the number represented by $(4, 5)$, we start at $(2, 3)$, and draw an arrow that represents moving 4 units in the horizontal direction (that is the direction of the *real axis*), and 5 units up in the direction of the vertical (the *imaginary axis)*. We end up at the point with coordinates $(6, 8)$. In much the same way we can define subtraction of complex numbers by subtracting the real and imaginary parts so that, for example, $(5 + 7i) - (1 + 2i) = 4 + 5i$. This can be pictured as starting with the vector $(5, 7)$, and subtracting the vector $(1, 2)$, to finish at the point $(4, 5)$. The next diagram (Fig. 10.1) illustrates examples of addition in the plane of the complex numbers.

Multiplication is quite another matter. Formally it is easy to do: we multiply two complex numbers together by multiplying out the brackets and remembering that $i^2 = -1$.* If we do this, we can even write down a rule on how to multiply the pairs together: $(a, b)(c, d) = (ac - bd, ad + bc)$.[51] Although a succint

---

51 The first to present multiplication of complex numbers explicitly in this fashion was William Rowan Hamilton in a paper to the Irish Academy in 1833.*

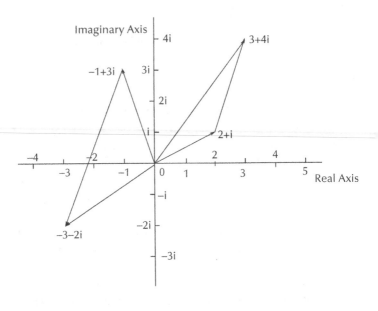

$(2+i)+(1+3i)=3+4i$ $(-1+3i)+(-2-5i)=-3-2i$

Figure 10.1. Sums in the complex plane.

enough rule, it can nonetheless leave you cold, for it looks quite awkward and meaningless. Before mentally retreating however, we should appreciate that the same kind of comment applies to the arithmetic of fractions. Let us take two ordinary fractions, $\frac{a}{b}$ and $\frac{c}{d}$ but, for the sake of comparison, let us consider them written as the ordered pairs (a, b) and (c, d) and look at the rules for combining them under addition, and under multiplication.

On this occasion we see that the rule for multiplication looks very natural and easy, while that for addition is relatively complicated. For multiplication we have: $(a, b)(c, d) = (ac, bd)$, while putting the two fractions over the common denominator of bd and adding gives: $(a, b) + (c, d) = (ad + bc, bd)$. This rule is only digestible after considerable experience with adding fractions together, for only then is it possible to see that the rule encapsulates what happens. However as long as a student understands how to add fractions, there is no need to memorize it.

The same applies with the rule for multiplying complex numbers—as long as a student can safely multiply out brackets, the rule of combination need not be committed to memory. However, we still lack a natural interpretation of complex multiplication that might easily be brought to mind and give it meaning.

The Polar System

Enlightenment comes through changing our perspective. Multiplication takes on a more meaningful form if we alter our coordinate system of the complex plane from the ordinary rectangular or cartesian coordinates as they are called to *polar coordinates*.[52]

In this system a point z is once again specified by an ordered pair of numbers, which we shall write as (r, θ). The number r is the *distance* of our point z from the origin O, (called in this context *the pole*). Therefore r is a non-negative quantity and all points with the same value of r form a circle of radius r centered at the pole.

---

52 Isaac Newton (1642–1727) described eight alternative coordinate systems for points in the plane, the seventh of which was the system of polar coordinates.

We use the second coordinate θ to specify z on this circle by taking θ to be the *angle*, measured in an anti-clockwise direction, of the line Oz from the real axis as shown in Fig. 10.2. The number r is called the *modulus* (plural moduli) of z, and we shall refer to θ as the *angle* of z.[53]

One slight blemish of the system is the exceptional nature of the pole itself, for its polar coordinates are not unique: no matter what value of θ we take, the point $(0, \theta)$ represents the origin, O.

Suppose now that we have two complex numbers, z and w, whose polar coordinates are (r_1, θ_1) and (r_2, θ_2) respectively. What are the polar coordinates of their product zw?

The rule of combination can be expressed neatly in ordinary language: the modulus of the product zw, is the product of the moduli of z and w, while the angle of zw is the *sum* of the angles of z and w. In symbols, zw has polar coordinates $(r_1 r_2, \theta_1 + \theta_2)$.*

The multiplication of the real numbers is happily subsumed under this more general way of looking at things: a positive real number r for instance has polar coordinates $(r, 0)$, and if we multiply by another, $(s, 0)$, the result is the expected $(rs, 0)$, corresponding to the real number rs.

Much more of the character of the multiplication of complex numbers however is revealed through this interpretation. The polar coordinates of the complex unit i are given by $(1, 90°)$.[54] If we now take any complex number $z = (r, \theta)$ and multiply by $i = (1, 90°)$, we find that $zi = (r, \theta + 90°)$. In other words

---

53 The standard word for this however is the *argument* of z.

54 Generally speaking, angles are not measured in degrees in such circumstances, but in the natural mathematical unit of the *radian*: there are 2π radians in a circle, so that a turn of one radian corresponds to moving one unit along the circumference of the unit circle, centered at the origin. One radian is just over 57°.

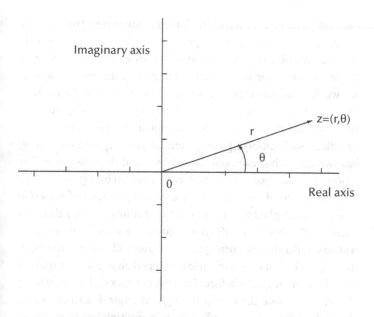

Figure 10.2. Polar coordinates of a point in
the complex plane.

multiplication by i corresponds to a rotation through a right angle
about the center of the complex plane.

Pythagoras, of course, had no inkling of any of this but he
certainly would have appreciated the significance of this revelation:
the right angle, that most fundamental of geometric ideas, can be
represented by a number.

Indeed the effect of adding or multiplying by a complex number
z on all the points in a given region of the complex plane can

each be pictured geometrically. Imagine any region you fancy in the plane. If we *add z* to every point inside your region we simply move each point the same distance and direction determined by the arrow, or vector as we often call it, represented by z. That is to say we translate the region to some other position in the plane so that the shape and size are exactly maintained, as is its attitude, by which we mean the region has not undergone any rotation or reflection. Multiplying every point in your region by $z = (r, \theta)$ has two effects however, one caused by r and the other by θ. The modulus of each point in the region is increased by a factor r, so all the dimensions of the region are increased by a factor of r also (so its area is multiplied by a factor of r^2). Of course if $r < 1$ then this 'expansion' is better described as a contraction as the new region will be smaller than the original. The region will however maintain its shape—for instance, a triangle is mapped on to a similar triangle with the same angles as before. The effect of θ, as we have explained above, is to rotate the region through an angle θ, anti-clockwise about the origin. The net effect then in multiplying all points of your region by z is to expand and rotate your region about the pole. The new region will still have the same shape as before but will be a different size and it will be lying in a different attitude determined by the rotation angle θ.

Gaussian Integers

The complex numbers $z = a + bi$, where a and b are themselves ordinary integers, form a lattice pattern in the complex plane, and are known as the *gaussian integers*. A gaussian integer is *prime* if it cannot be factored as a product of other gaussian integers (excepting the units $\pm 1, \pm i$). Happily there is an analagous result to the Fundamental Theorem of Arithmetic that guarantees that the prime factorization of an ordinary integer is unique. Any gaussian integer can be factorized uniquely as a product of powers of i (which are i, $i^2 = -1$, $i^3 = -i$ and $i^4 = 1$) and the *positive* gaussian primes, which are those gaussian primes whose real part is at least as great as their imaginary part.

Many ordinary prime integers are no longer prime when viewed as gaussian integers. For example, we must expunge both 2 and 5 from the list as $2 = (1 + i)(1 - i)$ and $5 = (2 + i)(2 - i)$. Indeed any positive integer n that is the sum of two squares cannot be a gaussian prime because then we obtain

$$n = a^2 + b^2 = (a + bi)(a - bi)$$

Gaussian primes can be used to show that an ordinary odd prime number is the sum of two squares if and only if it has the form $4n + 1$. This is a stepping stone in the characterization of numbers that are the sum of two squares. A number will have this property unless its prime factorization contains a prime of the form $4n - 1$, raised to an odd power. The use of guassian primes allow us to find all the ways a number can be written as the sum of two squares and so allows us to count how many such representations there are.

The complex number approach gives a natural way of seeing that the product of two numbers that are each the sum of two squares is another of the same kind. For instance, let $x = a^2 + b^2$ and $y = c^2 + d^2$ and consider the related gaussian integers $z = a + bi$ and $w = c + di$. The *complex conjugate* of a complex number z is the number $\bar{z} = a - bi$ that results by reflecting z in the real axis, and as we saw above $z\bar{z} = a^2 + b^2$. One remarkable but easily verified property of the complex conjugate is that the product of conjugates is the conjugate of the product: in symbols $xy = \bar{z}\bar{w} = \overline{zw}$. Multiplying both sides of this equality by zw and rearranging the order slightly gives $(z\bar{z})(w\bar{w}) = zw\overline{zw}$. In particular this says that the product of two sums of squares is another sum of squares. Indeed, remembering that $zw = (ac - bd) + i(ad + bc)$ gives us an explicit formula:

$$(a^2 + b^2)(c^2 + d^2) = (ac - bd)^2 + (ad + bc)^2.$$

For example, $29 = 2^2 + 5^2$ and $52 = 4^2 + 6^2$. The above formula allows us to write $1,508 = 29 \times 52$ as the sum of two squares. In this instance we have $a = 2$, $b = 5$, $c = 4$, and $d = 6$ and so the right hand side of our formula gives $(8 - 30)^2 + (12 + 20)^2$, so that $1,508 = 22^2 + 32^2$.

To be sure, the above identity is readily verified without recourse to complex conjugate manipulation but the use of the gaussian integers gives a natural path to the result and hints that there may be generalizations that lie beyond.

A classical result, with some particularly remarkable proofs, is that any positive integer n is the sum of *four* squares.

Glimpses of Further Consequences

The polar version of complex numbers is particularly suited to the taking of powers and roots for to raise $z = (r, \theta)$ to some positive power n, we simply raise the modulus to that power, and add θ to itself n times, to give $z^n = (r^n, n\theta)$. The same formula applies to fractional and negative powers, and goes by the name of De Moivre's (1667–1754) Theorem.*

Division of complex numbers is something that we have neglected mentioning up until this point. As with real numbers, division by a complex number z means multiplication by its reciprocal $w = \frac{1}{z}$, but what number is this w? Given that $z = (r, \theta)$ the number w is the one with the property that $zw = (1, 0)$, the number 1. This shows us that we must take $w = (\frac{1}{r}, -\theta)$, for then $zw = (r, \theta)(\frac{1}{r}, -\theta) = (r\frac{1}{r}, \theta - \theta) = (1, 0)$, just as we require. Reciprocals and division can also be explained in rectangular coordinates using the idea of the so-called complex conjugate.*

In summary, addition and subtraction of complex numbers are easily interpreted using rectangular coordinates while multiplication, division, powers, and roots become more transparent when we approach the complex plane using polar coordinates.

There are a host of applications of complex numbers, even at the elementary level. The interplay between rectangular and polar representations bring trigonometry into play in a surprising and advantageous way. For instance, a standard exercise for students is the derivation of important identities that now arise in a very natural way, by taking arbitrary complex numbers of unit modulus (i.e. $r = 1$), and calculating powers using both rectangular and then polar coordinates. Equating the two forms of the answer then reveals a trigonometric equation.

Things quickly get deeper. One of the most celebrated formulas in mathematics appeared in 1748 in the textbook *Introduction to infinite analysis* in which Euler deduced the stunning little equation, $e^{i\pi} = -1$, relating the four most mysterious numbers in the world. Indeed Euler's equation eventually became a key ingredient in a proof of the impossibility of squaring the circle (1882). The equation features in the proof that π is a transcendental number, one that is not the solution of any polynomial equation with rational coefficients. In consequence, the age old problem of whether a square can be constructed with straightedge and compasses with area equal to that of a given circle is answered in the negative. It cannot be done.

And much more was to come: the use of complex numbers reveals a connection between the exponential, or power function and the seemingly unrelated trigonometric functions. Without passing through the portal offered by the square root of minus one, the connection may be glimpsed, but not understood. The so-called hyperbolic functions arise from taking what are known as the even and odd parts of the exponential function.* To every trigonometric identity there corresponds one of identical form, except perhaps for sign, involving these hyperbolic functions. This can be verified easily in any particular case, but begs the question as to why it should happen at all.[55] Why should the behavior of one class of functions be so closely mirrored in another class, defined in so different a manner, and of such different character?

The hyperbolic functions can be introduced geometrically by analogy with the trigonometric functions, through replacing the underlying circle by which the trigonometric functions are introduced, by a different curve known as a hyperbola, and this accounts for their name. This geometric link does not however explain the closely matching pattern. Resolution of the mystery is by way of the formula first enuciated by Euler that $e^{i\theta} = \cos\theta + i\sin\theta$, which shows that the exponential and trigonometric functions are intimately linked, but only through use of the imaginary unit i.[56] Once

55 The exact nature of the correspondence is governed by what is known as Osborne's Rule.*
56 Others, such as Jean Bernoulli (1667–1748), also were familiar with this relationship.

this is spotted, it becomes clear that results along the lines above are inevitable, by performing calculations using the two alternative representations offered by Euler's equation and then equating real and imaginary parts. Without the formula however, it all remains a mystery.

In the 19th century, the theory of a function of a complex variable was founded as a subject in its own right by Augustin Cauchy (1789–1857). Cauchy himself was in many ways a strange man. The most charitable description Bertrand Russell could offer of him was that he had very peculiar principles but such as they were he lived by them. He was though a very great and prolific mathematician, and was the founder of the theory of complex variables that is now one of the pillars of mathematics.

The arithmetic of the complex numbers itself reveals many surprises, a few of which have been referred to above, but by re-working the whole of the calculus with complex, as opposed to real variables, a new mathematical world swung into view. For example, one of the first rewards the theory has to offer is Cauchy's Residue Theorem. This comes as a complete surprise and is an extraordinary and powerful result. Intractable questions involving areas of curves defined by ordinary real variables suddenly become accessible by allowing your variable to leave the monorail that is the real line and roam freely over the entire complex plane. The nature of things often only becomes clear when we take this wider view.

The applications of the theory of complex variables is too immense to do it justice here. For example, the analysis of matter on an atomic scale is studied through x-ray diffraction—the underlying picture is recovered from the manner in which electromagnetic waves are scattered when they encounter the object.

The recovery process depends on the workings of so-called Fourier transforms in which it is crucial that the variables involved are complex, and not just real numbers. The entire subject rests upon this technique—complex numbers are not just a mathematical abstraction but are 'real' and they work!

The beauty of the complex plane is that we may finally carry out all our mathematical work in a single number arena. However, although there may be no pressing mathematical difficulty that is driving us further, we can ask the question whether or not it is possible to go beyond the complex plane into some larger realm of number. After all, we now have a number system based on pairs of real numbers, two-dimensional vectors if you like. It is natural to ask therefore is there some way of developing a number system based on triples of numbers, which contains the working of the complex numbers in the first two coordinates, just as the system of complex numbers has a copy of the real numbers embedded through the first members of the pairs. The answer is very surprising indeed. It can't be done in three dimensions, but it can be done with four.

For ten years, William Rowan Hamilton (1805–1865) mused on the problem of developing a number system based on triples of the form $a + bi + cj$, where $a + bi$ was a complex number, c was real and j was some new kind of unit. While out walking with his wife, he had a flash of inspiration. An extension could be made to work on expressions of the form $a + bi + cj + dk$. However, the multiplication could no longer be commutative. In fact the rules of the game had to be that $i^2 = j^2 = k^2 = -1$, $ij = k$ but $ji = -k$ with similar rules for other products. So impressed was he by his own genius that he scratched the defining equations as graffiti on

Brougham Bridge over the Royal Canal in Dublin. On October 16th 1843, the *quaternions* were born.

Hamilton remained enraptured by his discovery for the rest of his life. He quickly found that the appropriate measure of size, or *norm* as it is called, of one of his quaternion numbers is the square root of the sum of the squares $a^2 + b^2 + c^2 + d^2$, for the norm of the product of two quaternions is the product of the norms. This is equivalent to an observation of Euler that if two integers are each the sum of four squares then so is their product. (The first step in the proof that *every* number is the sum of four squares, for it reduces the question to that of solving the same problem for the primes alone.) The corresponding result for two squares we deduced in the section on gaussian integers by making use of (the square of) the norm of a complex number z in the form $z\bar{z}$. However, up until the discovery of the quaternion norm, Euler's identity was a rabbit-out-of-a-hat.[57] Now Hamilton could give it a meaningful interpretation, a natural path to its proof, which no doubt convinced him further of the value of his big new idea (For an overview see Fig. 10.3.).

It has transpired however that the quaternions, although a generalisation of the complex numbers, do not seem to be as important. Nonetheless their discovery had a tremendous galvanising effect on 19th century mathematics, for it showed that a consistent algebra could be built that satisfied most, but not all the usual Laws of Algbera. These Laws therefore were not as immutable as everyone had supposed. Mathematics had been given

---

57 The number n can be written as the sum of *three* squares unless it has the form, $4^e(8k + 7)$; for example 7 is not the sum of three squares*.

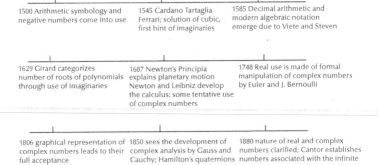

Figure 10.3. Time line of the use of number
in European mathematics.

a new freedom to explore new algebraic systems and the theory of
matrices, a very different kind of numerical object, was now set
to flourish. One hundred and fifty years on, the theory of linear
algebra, which has matrices as its primary objects, is among the
most applied topics in all of mathematics.

In 1867, Hankel proved that the algebra of complex numbers is
the most general possible that obeys *all* the laws of ordinary arithmetic. Indeed there are severe limitations in the quest for making
vectors strings of length more than two behave like an extension
of the complex numbers. Apart from Hamilton's quaternion system, there is no other where general division is possible except
for Cayley's *octonians,* which fail associativity of multiplication, so
that the bracketing of products in different ways yields different
outcomes. In summary, it is possible to go beyond the system of
the complex numbers, but the nature of the algebras that arise

are less structured and, it seems, generally less important.[58] The plane of the complex numbers will always remain one of the central discoveries of mathematics. But we may still ask...

Can people just keep making up new types of numbers?

The number-like systems have certainly expanded over the centuries. First there were just counting or natural numbers 1, 2, 3,...then, in some order, came fractions, zero, negatives of whole numbers and fractions, forming what we now call the set of rational numbers. However, even from the days of Pythagoras, fractions were not enough to describe all numerical phenomena for, as mentioned above, $\sqrt{2}$ is not a fraction. This has led to the *real* numbers, which we may think of as the collection of all possible decimal expansions. However the nature of mathematics itself has led us, at first reluctantly, to go beyond real numbers to the realm of the so-called imaginary and complex numbers. Moreover modern mathematicians also deal in infinite numbers of more than one kind, and also quaternions, octonians, and matrices, which can be regarded as another generalization of number.

This proliferation of number types may leave the false impression that mathematicians spend their time whimsically inventing new numbers for no good reason. This is not the case. In each instance, the new extended number system incorporates the original number systems within in such a way that most, if not all of the usual Laws of Algebra persist. This expectation places severe restraints on the possibilities of new number types. What is more, the plane of the complex numbers has turned out to be such a

---

58 This direction of generalization is by no means barren: for example so called Clifford algebras, which are important in sub-atomic physics, are a development along these lines.

natural arena for number work that it has largely obviated the need to go beyond it.

As a comparison, it is always possible to invent a new language, be it spoken or for programming on a computer, but no-one will be interested in your new tongue unless it allows you to do something better, or faster, or it lets you express and understand interesting things in an enlightening way. All the new number types mentioned above satisfy these criteria, which is why they have become a part of modern mathematics. New number types of genuine interest do arise from time to time, but not at all often.

The Number Line under the Microscope

In Chapter 7 we saw how the real number line was a densely packed mix of the rational and the irrational. If the rational points were blue and the irrational red, what would we see? There would be a red dot between every pair of blue dots, and a blue dot separating every pair of reds, so we might expect the overall effect to be one of uniform purple. On the other hand, the blue points form only a countable set, which has measure zero, compared with the remaining red points, so the effect of the red would surely swamp that of the blue making the latter invisible. Neither interpretation stands up to scrutiny as there is no physical experiment that could approximate the limiting behavior of which we are speaking. We need to think of the line more in its own terms.

Whatever the structure of the real line, it surely consists of copies of the unit interval $I = [0, 1]$ repeated over and over.

(*I* consists of that part of the number line from 0 to 1 inclusive.) If we could describe how our numbers sit within this interval therefore, we would have a complete picture of how the structure of the entire line is knitted together, as it consists of uniform repetition of the basic interval *I*.

The most accessible numbers in *I* are the rationals, and among them those with small denominators (and so small numerators as well) are the most special. We can ask how those numbers sit within the interval. We therefore choose a small counting number, *n*, and look at all the fractions in *I* that can be written using numbers no larger than *n*. For instance, we might try $n = 7$, and list the corresponding sequence of fractions in ascending order:

$$\frac{0}{1}, \frac{1}{7}, \frac{1}{6}, \frac{1}{5}, \frac{1}{4}, \frac{2}{7}, \frac{1}{3}, \frac{2}{5}, \frac{3}{7}, \frac{1}{2}, \frac{4}{7}, \frac{3}{5}, \frac{2}{3}, \frac{5}{7}, \frac{3}{4}, \frac{4}{5}, \frac{5}{6}, \frac{6}{7}, \frac{1}{1}.$$

This sequence is called the seventh *Farey sequence* of fractions. The Farey sequences are alive with surprising algebraic and even geometrical properties. For example, each term in the sequence can be got by adding the numerators and denominators of those on either side: applying this for example to $\frac{1}{6}$, we compute $\frac{1+1}{7+5} = \frac{2}{12}$, which cancels to $\frac{1}{6}$; similarly if we look at the neighbors of $\frac{3}{4}$, we get $\frac{5+4}{7+5} = \frac{9}{12}$, which cancels down as it should. Another pattern to note is that if we take the difference when we cross multiply in two successive fractions, the answer is always 1: for example take, $\frac{2}{5}$ and $\frac{3}{7}$: the cross multiples are $2 \times 7 = 14$ and $3 \times 5 = 15$, and they differ by 1.[59]

59 The sequence is named after Farey who wrote an article on the subject in which he stated the first of these properties much as we have done, without proof. It seems though that both these results were first stated and proved by Haros in 1802, some 14 years before Farey's article.

On the other hand, it is not so obvious how we go about writing down the sequence: for any given n we can write down all the fractions in the nth Farey sequence (although due to the erratic nature of cancellation, it is not clear how many of them there will be*) and, by comparison, eventually arrange them in order. However there is a better way of doing it: given one fraction in the sequence it is possible to calculate directly what the next one will be, but it is not a trivial matter (see Hardy and Wright, *An Introduction to the Theory of Numbers*).

We can investigate the placement of an individual irrational number a in I by asking how closely it is settled to the Farey sequences. Of course a will never lie in any Farey sequence F_n, but it could happen that some irrationals a nestle much closer to members of F_n than others.

To explain further, it is of course possible to approximate any irrational a as closely as we please by rationals, as this is what we do when we take the decimal expansion of a, and truncate it further and further along the expansion. This gives a sequence of rational numbers that march every closer to the given number a, which is represented by the expansion taken in its entirety. However, these rationals, when expressed as vulgar fractions, may have extremely large denominators, in which case in order to get very close to a we would have to take fractions from Farey sequences with very high values of n.

Since F_n contains all the numbers $\frac{m}{n}$, as m ranges from 0 up to n, it follows that every number in I can be approximated by a member of F_n, to an accuracy of $\frac{1}{n}$. However it can be shown, and we shall outline how this comes about, that *for every* irrational number a in the unit interval, there are infinitely many values of n, for which some member of F_n differs from a by less than $\frac{1}{n^2}$. That is to say there are larger and larger values of n for which members of F_n comes very close to a indeed.

This is true for any irrational number at all. However some irrational numbers can be consistently appoximated much better than this. Which irrationals can be well approximated by rationals in this way, and which cannot? The simple algebraic numbers, like $\sqrt{2}$, seem closest in nature to the rationals, while we might expect that non-algebraic numbers, the transcedentals, to live apart and not to have close rational neighbors. Surprisingly, the opposite is true. On the one hand, it can be proved that any irrational number that can be well-approximated by rationals (in a sense that can be made precise) must be transcendental. Indeed this affords one of the standard techniques for showing that a number is transcendental. From the point of view of rational approximation, it is the simplest irrational numbers that are the worst. Numbers like $\sqrt{2}$ and those related to the *Golden Ratio*, $\frac{1}{2}(1 + \sqrt{5})$, are the hardest to approximate of all. To see why is quite some little story in itself, which begins again with unit fractions.

Return to Egypt

If we are still hanker after the Egyptian aesthetic, and prefer fractions with numerators of unity, we might be tempted to ask is there any way we can turn other vulgar fractions into ones very like them. We could begin with a fraction such as $\frac{2}{7}$, and divide top and bottom by 2, to give $\frac{1}{3+\frac{1}{2}}$. This does give us a single fraction of sorts in which all numerators are 1. Another example

$$\frac{25}{91} = \cfrac{1}{3 + \frac{16}{25}} = \cfrac{1}{3 + \cfrac{1}{1 + \frac{9}{16}}} = \cfrac{1}{3 + \cfrac{1}{1 + \cfrac{1}{1 + \frac{7}{9}}}} = \cfrac{1}{3 + \cfrac{1}{1 + \cfrac{1}{1 + \cfrac{1}{1 + \frac{2}{7}}}}}$$

$$= \cdots = \cfrac{1}{3 + \cfrac{1}{1 + \cfrac{1}{1 + \cfrac{1}{3 + \frac{1}{2}}}}}$$

It seems pretty clear that it can always be done, but, at the same time, you would think that not even the most obsessed Egyptian fraction zealot would claim any practical worth for this calculation. If you pursue a few more examples however, one rather neat feature emerges all the same. We began with a *reduced fraction*, one that was cancelled to its lowest terms, and all the intermediate fractions in the calculation were also similarly reduced. This happens every time. What would happen if we *egyptianised* a fraction that was not reduced?

$$\frac{84}{105} = \cfrac{1}{1 + \frac{21}{84}} = \cfrac{1}{1 + \frac{1}{4}},$$

$$\frac{2058}{3675} = \cfrac{1}{1 + \frac{1617}{2058}} = \cfrac{1}{1 + \cfrac{1}{1 + \frac{441}{1617}}} = \cfrac{1}{1 + \cfrac{1}{1 + \cfrac{1}{3 + \frac{294}{441}}}} = \cfrac{1}{1 + \cfrac{1}{1 + \cfrac{1}{3 + \cfrac{1}{1 + \frac{147}{294}}}}}$$

$$= \cfrac{1}{1 + \cfrac{1}{1 + \cfrac{1}{3 + \cfrac{1}{1 + \frac{1}{2}}}}}$$

The highest common factors of numerator and denominator in each of these cases is respectively 21 and 147, which are the numerators that turned up in the penultimate step of the calculation. It seems that the highest common factor of two numbers can be found by egyptianising the corresponding fraction. Why should this be so?

What is working for us here is what is known as the euclidean algorithm, a simple idea that is nevertheless one of the most important in algebra. This in turn rests upon one very simple observation. Suppose that we subtract b from a, to leave a remainder r: $a - b = r$. Then *any common factor of a and b is also a factor of r.*[60] Indeed any number that is a factor of two of these numbers must also be a factor of the third. In particular, the highest common factor (hcf) of a and b is also the highest common factor of b and r. Since b and r are smaller than a, it is easier to continue to work with that pair instead in the hunt for the hcf and we now repeat the process: let $b - r = s$, and work on with the pair r and s. Since the numbers involved are all positive, and decreasing, this process must eventually cease when the two numbers in hand, u and v say, are equal (so that $u - v = 0$, and we can go no further). Clearly the hcf of u with itself is u, so that u is the hcf of the original pair of numbers, a and b. This is the euclidean algorithm for finding the highest common factor of two given numbers. It allows us to determine it without factoring the numbers a and b, which is very important, as it takes much more work to factor numbers than it does to subtract them.

If we apply the algorithm to the pair of numbers $(3675, 2058)$ for example, the number pairs that we obtain run as follows:

$$(3675, 2058) \rightarrow (2058, 1617) \rightarrow \underline{(1617, 441)} \rightarrow (1176, 441) \rightarrow$$
$$\underline{(735, 441)} \rightarrow (441, 295) \rightarrow (294, 147) \rightarrow (147, 147),$$

so 147 is the hcf of 3675 and 2058.

---

60 Let c be such a common factor so that $a = cd$ and $b = ce$ say. Then $r = a - b = cd - ce = c(d - e)$, and so r is also a multiple of c.

We see all this reflected in the egyptian calculation of the corresponding fraction. The point to note relates to the underlined pairs, where the smaller number of the pair, 441, features in three successive pairs. This corresponds to the 3 in the line of the fraction at this point, and arises because the intermediate remainder, 441 is small enough compared to 1617 that it may be subtracted more than once, three times in fact, and so we do so. In practice, this is how the euclidean algorithm runs. For example, to find the hcf of 224 and 98, application of the euclidean algorithm and the corresponding egyptianisation of the fraction look like this:

$$224 = 2 \times 98 + 28$$

$$98 = 3 \times 28 + 14$$

$$28 = 2 \times 14$$

$$\frac{98}{224} = \frac{1}{2 + \frac{28}{98}} = \frac{1}{2 + \frac{1}{3 + \frac{14}{28}}} = \frac{1}{2 + \frac{1}{3 + \frac{1}{2}}}$$

We see that the hcf in this instance is 14.

The standard term for this fractional realization of the number is its *continued fraction*. We see that there is one line in the continued fraction for every line of the euclidean algorithm when performed on the two numbers. In particular, starting with a reduced fraction in which the two numbers have a hcf of 1 (we say the numbers are *relatively prime* in these circumstances as they have no common prime factor) the same will be true of all the fractions that arise in the course of the calculation of the corresponding continued fraction. We'll return to this idea after a short digression.

Coin Problems, Sums, and Differences

Juggling and pouring problems may seem more in the realm of riddles than serious mathematics. There is an apocryphal connection with mathematics however dating back to the 19th century. The story goes that the eminent French mathematician, Siméon Poisson, was a complete professional failure until he came across a problem like our next one. While others around him in the train carriage got muddled, he had no trouble finding the solution, so coming to realize that he had a talent that might be put to use after all. Here is Poisson's Problem.

Two friends have an eight litre jug full of wine that they wish to share evenly. They have two empty vessels of capacities three and five litres respectively. The question is, how can they share their wine fairly?

Two four-litre portions can be created in seven steps. You may compare your own efforts with that of Poisson's solution here (Fig. 11.1). The initial situation is written as $(8, 0, 0)$, indicating the respective amounts held in the eight, five, and three litre jugs.

The evolution of the solution then goes by way of the following stages—the way to pass from each stage to its successor being clear enough:

$$(8, 0, 0) \to (3, 5, 0) \to (3, 2, 3) \to (6, 2, 0) \to (6, 0, 2)$$
$$\to (1, 5, 2) \to (1, 4, 3) \to (4, 4, 0).$$

Another problem of a similar type, involves sums rather than differences. It is more convenient to revert for the moment to English coinage. What sums are possible using only two and five pence pieces?

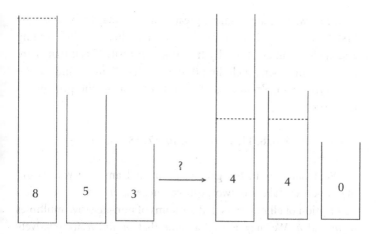

Figure 11.1. Poisson's juggling problem.

Since four and five pence sums are possible, and any number greater than three can be obtained by starting at either four or five pence and adding a suitable number of 2p pieces, it follows that all sums are feasible with the obvious exceptions of one and three pence. Let us try a more challenging example of a *coin problem* as it is known.

What numbers can be expressed as sums of multiples of 3 and 8?

A little experimentation leads one to discover the following line of attack. The smaller number is 3, so try to locate three successive numbers that are feasible. You will find the smallest such trio is 14, 15, and 16:

$$1\,4 = 8 + 2 \times 3,\ 15 = 5 \times 3,\ 16 = 2 \times 8.$$

It follows that any number greater than 13 may be obtained by first creating 14, 15, or 16 as the case may be, and then adding a suitable number of 3's. By trial one can easily check that there are seven numbers, including 0, that are less than 14 and can be generated from 3's and 8's. The full solution to the problem is therefore

$$0, 3, 6, 8, 11, 12, 14, 15, 16, 17, 18, 19, 20, \cdots$$

What happens in the general problem I am sure you would like to know. Take any two positive integers, m and n. We seek a description of all numbers that are sums of non-negative multiples of m and n. We may as well assume that m and n are relatively prime, meaning that they have a highest common factor of 1, for if their hcf is d, clearly we can only create multiples of d. For example, if we used as generating numbers 6 and 16 instead of 3 and 8, the answer would be found by taking the above solution for the pair 3 and 8 and doubling all the numbers. In general we would work with the pairs $\frac{m}{d}$ and $\frac{n}{d}$, which have an hcf of 1, solve the problem for this case, and multiply each of the resulting numbers by d to find the answer for the original pair of generators m and n. In other words, we would work in units of d instead of units of 1.

When we take m and n to be relatively prime we find that the answer follows the pattern set by our examples above. All numbers can be generated from the value $(m-1)(n-1)$ onwards, and the number previous to that is impossible. In our problem, $(m-1)(n-1) = 2 \times 7 = 14$, and 13 was not feasible. What is more, *exactly half* of the numbers from 0 up to this final forbidden value can also be generated. This is also consistent with our example where we saw that 7 of the 14 numbers $0, 1, 2, \cdots, 13$ were possible using the generators 3 and 8.

There are situations where we are left wondering as to the uniqueness of the decomposition, and one scenario comes about in scores in sporting encounters. In particular, in American grid iron football, most team scores are of the form $3m + 7n$. At least they are if we assume no two-point conversions of touchdowns, no missed extra points, and no safeties. For those of you unfamiliar with this jargon, all you need to know is that scoring is nearly always done in multiples of 3 (field goals) and 7 (converted touchdowns.) Our theory above tells us that, with these restrictions, any score from $2 \times 6 = 12$ upwards is possible, but not a score of 11 points. However, in some cases there is only one way the score can be arrived at. For instance, when Indianapolis beat Baltimore in the 2007 playoffs by a score of 15 to 6, it means that it was a game consisting entirely of field goals, as there was no other way of reaching those numbers by way of 7 and 3 point scores. That is to say, you can, to some extent, decompose the game just from the final score. It tells the seasoned football fan enough for him to know just what kind of a game it must have been without even having to watch it!

The two-coin problem represents the first and last reasonably easy case of this problem type—the same question where you allow yourself three coin denominations is much harder.[61] In the mathematical literature, the first number that cannot be generated in this way, $mn - m - n$ in the two-coin case, is called the *Frobenius number* of the corresponding numerical semigroup.

---

61 The two-coin problem was first solved by Sylvester in 1884. An explicit solution has been found for the three-coin problem, but the general n−coin problem is known to be of a type that is particularly intractable—the class of NP hard problems.

However, the number generating problem is both more important and easier to handle if we allow general integer multiples and do not insist on non-negative multipliers only, and here the link emerges with the euclidean algorithm.

The question is, which integers are representable as a sum of multiples of two given integers, m and n? Once again it is clear that only multiples of d, the highest common factor of m and n, will be feasible. What is more, if we can express d in the form $am + bn$, then *all* multiples of d will be possible as well—to get kd we would simply take $(ka)m + (kb)n$. The question is, can we get our d?

The answer is yes, and the method is to work the euclidean algorithm backwards. For example, let us take $m = 3$ and $n = 8$. Applying the algorithm we find

$$8 = 2 \cdot 3 + 2$$
$$3 = 1 \cdot 2 + 1$$
$$2 = 2 \cdot 1$$

confirming that the hcf of our two numbers is 1. The approach is to focus on the second to last line of the algorithm, where the hcf first appears. Use that equation to write the hcf, 1 in this case, in terms of the previous number pair, and continue to use each equation to eliminate the intermediate numbers until the hcf is written in terms of the original pair. In this case we obtain:

$$1 = 3 - 1 \cdot 2 = 3 - 1 \cdot (8 - 2 \cdot 3) = 3 \cdot 3 - 1 \cdot 8$$

A similar example the reader might care to work through is with the pair $(516, 432)$. The euclidean algorithm yields that the hcf is 12, and working the equations backwards gives $12 = 6 \times 432 - 5 \times 516$.

The fact the the hcf can always be expressed in this way is of immense theoretical importance. It can be used for example to prove *Euclid's Lemma,* which assures that if a prime number p is a factor of a product ab, then p is a factor of at least one of the numbers a and b individually.* This in turn is the key to proving the Fundamental Theorem of Arithmetic that says that there is one and only one way to factorize a number into a product of primes.

Fibonacci and Fractions

Recall the sequence of numbers, 1, 1, 2, 3, 5, 8, 13, 21, $\cdots$ discovered by Fibonacci and introduced in Chapter 4. If we look at the continued fraction representation of a pair of successive Fibonacci numbers a strikingly simple fact emerges. Take for instance

$$\frac{13}{8} = 1 + \frac{5}{8} = 1 + \frac{1}{1 + \frac{3}{5}} = 1 + \frac{1}{1 + \frac{1}{1 + \frac{2}{3}}} = 1 + \frac{1}{1 + \frac{1}{1 + \frac{1}{1 + \frac{1}{1}}}}$$

We obtain a long fraction consisting entirely of 1's, and each preceding ratio of Fibonacci numbers appears as we wind through the calculation. This must happen every time: by the very way they are defined, each Fibonacci number is less than twice the next, and so the result of the division will leave a quotient of 1 and the remainder is the preceding Fibonacci number. You will recall that the ratio of successive Fibonacci numbers approaches the Golden Ratio, τ, and so this suggests that τ is the limiting value of the continued fraction consisting entirely of 1's.

This can be confirmed in a rather clever way. If we call the value of the infinite fractional tower of 1's by the name a, we see that a

satisfies the relation $a = 1 + \frac{1}{a}$, because what lies underneath the first floor of the fraction is just another copy of a. From this we see that a satisfies the quadratic equation $a^2 = a + 1$, the positive root of which is $\tau = \frac{1+\sqrt{5}}{2}$. This proves the original observation of Kepler that ties the Fibonacci numbers to the Golden Ratio.[62]

What is more, it is the fact that the continued fraction version of τ contains nothing but 1's that makes it so very hard to approximate it by rational numbers. This example opens the door to the idea that we may be able to represent other irrational numbers not by finite continued fractions (which are obviously just rational themselves) but by infinite ones. Continued fractions look very awkward, partly because of the many floors we have used in representing them, but they are of genuine mathematical significance, as we have seen already as the pattern of their behavior gave us a path to the euclidean algorithm. We can however form a continued fraction for any number $a > 1$ in much the same way as we do for rational numbers.* The inconvenience of writing all the floors of the division is easily side-stepped—since all the numerators that we are using are 1's, we only need to record the quotients in the division to specify which continued fraction we are talking about. For instance the representation for the fraction $\frac{25}{91}$ is specified by the list $[0, 3, 1, 1, 1, 2]$ and the Golden Ratio, τ has the continued fraction representation $[1, 1, 1, 1, \cdots]$. In a fashion reminiscent of repeating decimal notation we write $\tau = [\overline{1}]$.

In this way we see that the irrational number τ has a recurring representation as a continued fraction. Indeed the numbers that have recurring representations as continued fractions are

62 First proved by the Scottish mathematician Robert Well in 1753.

rational numbers (which are exactly the ones whose representations terminate) and those that arise from quadratic equations such as τ, which we saw above is one solution of the equation $x^2 = x + 1$, and $\sqrt{2} = [1, \overline{2}]$, which satisfies $x^2 = 2$. Some other examples showing the rather unpredictable nature of the recurrences are $\sqrt{3} = [1, \overline{1, 2}]$, $\sqrt{7} = [2, \overline{1, 1, 1, 4}]$, $\sqrt{17} = [4, \overline{8}]$ and $\sqrt{28} = [5, \overline{3, 2, 3, 10}]$. There is nonetheless one very particular and remarkable facet to the pattern of the expansion of the continued fraction of an irrational square root. The expansion begins with an integer r, and the recurrent block consists of a palindromic sequence (a sequence of numbers that reads the same in reverse) followed by $2r$. This can be seen in all the preceding examples: for instance for $\sqrt{28}$ we see that $r = 5$, the palindromic part of the expansion is 3, 2, 3, which is followed by $2r = 10$. For $\sqrt{2}$ and $\sqrt{17}$, the palindromic part is empty, but the pattern is still there, albeit in a simple form. It can be shown that the continued fraction representation of a number is unique—two different continued fractions have different values.

The importance of continued fractions in approximation of irrationals by rationals is that the so called *convergents* of the fraction, which are the rational approximations of the original number that result from truncating the representation at some point and working out the corresponding rational number, are the best approximation possible in the sense that any better approximation will have a larger denominator than that of the convergents. The convergents of the Golden Ratio are the Fibonacci ratios. Since every term in the continued fraction representation of τ is 1, the convergence of these ratios is retarded as much as it possibly could be. For that reason there is no more difficult number than τ to approximate by rationals and the Fibonacci ratios are the best you can do.[63]

Some famous transcendentals have continued fraction representations that do not involve only unit denominators. The first is*

$$e = 2 + \cfrac{1}{1 + \cfrac{1}{2 + \cfrac{2}{3 + \cfrac{3}{4 + \cfrac{4}{5 \cdots}}}}}$$

Although transcendental, e is more akin to quadratic irrationals as it cannot be approximated by rationals any better than they. The first specific number to be proved to be transcendental through rational approximation is called Louiville's number. This number was not genuinely significant in itself, in that it was tailor made to fit an argument of Louiville, which showed that any number whose decimal expansion converged extremely swiftly (by consisting mainly of zeros with very sparse non zero entries that rapidly became rarer as we look along the expansion) must be transcendental, as any irrational algebraic number could not have convergents that were that efficient.

Another old favorite involving π is[64]

$$\frac{4}{\pi} = 1 + \cfrac{1^2}{2 + \cfrac{3^2}{2 + \cfrac{5^2}{2 + \cfrac{7^2}{2 + \cdots}}}}$$

63 If the denominator of a convergent is q, then the approximation is within $\frac{1}{\sqrt{5}q^2}$ of the true value of the number. The convergents of a continued fraction alternately underestimate and overestimate the value to which they approach.

64 This one is derived from the so-called Wallis product: an infinite product that is equal to π, found in the 17th century by the English mathematician John Wallis and is found through studying the areas under curves defined by successive powers of the sine function.

Cantor's Middle Third Set

Now we have more idea on the way the various sets of numbers that make up the number line sit amongst one another, it is time to return to the sizes of these sets. We have seen in a previous chapter that the rational numbers form a countable set, yet are densely packed into the number line. Cantor's Middle Third Set is by way of contrast an uncountable subset of the unit interval I, that is nevertheless sparsely spread.

We begin with the unit interval I, that is all the real numbers from 0 up to 1. The first step in the formation of Cantor's set is the removal of the middle third of this interval, that is all the numbers from $\frac{1}{3}$ to $\frac{2}{3}$ inclusive. The set that remains consists of the two intervals from 0 up to $\frac{1}{3}$ and from $\frac{2}{3}$ up to 1. At the second stage we remove the middle third of these two intervals, at the third stage we remove the middle third of the remaining intervals, and so on. Cantor's Middle Third Set is then all the points of I that are *not* removed at any stage of this process (Fig. 11.2).

What is the measure of this set C? We begin with the set C_0 consisting of the unit interval I, which has length one unit. At each stage the set remaining is two thirds the length of the set at the previous stage, since one third of it is discarded. Hence C_1 consists of two intervals of total length $\frac{2}{3}$, the next set consists of four intervals of length totalling $(\frac{2}{3})^2$, and in general, at the nth stage, we are left with a set C_n, which consists of 2^n little intervals whose total length is $(\frac{2}{3})^n$. We define the Cantor set as the set C, which is the set of points that are never discarded, that is to say are common to all the sets C_n. Now the sets C_n form a decreasing chain of ever smaller sets. Is there anything remaining that sits inside them all?

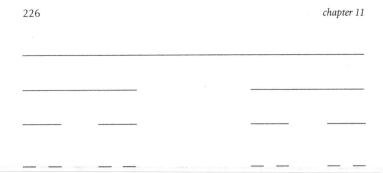

Figure 11.2. Evolution of Cantor's Middle
Third Set to the 4th stage.

Certainly the measure of the Cantor set is 0. Without introducing a formal notion of measure, we see that the set C is contained in a set of intervals whose collective length is $(\frac{2}{3})^n$. As n increases beyond all bounds the power of this fraction tends to 0. Therefore the Cantor set C cannot be assigned a positive measure $a > 0$, as C is contained inside sets of intervals whose collective length is less than a. The only value that can be given to the measure of the set C is zero. Like the set of ordinary fractions, the Cantor set therefore is a set in the unit interval I of measure 0.

We might suspect however that we have thrown the baby out with the bath water and that there are no points at all left in C. Is the Middle Third Set empty?

The answer is a resounding no! There are infinitely many numbers left in C.

To see this, it is easiest to shift to base three 'decimals' known as *ternary*, as the whole construction is based on thirds. In base 3 decimals the numbers $\frac{1}{3}$ and $\frac{2}{3}$ are respectively given by 0.1 and 0.2. By discarding the middle third of the unit interval we have thrown away all those numbers whose ternary expansion begins with 0.1. (We have also thrown out 0.2 which is equal to 0.1111 ...). At the second stage we have discarded all those numbers that have ternary expansions that begin 0.01 (the middle part of the first interval of one third), or 0.21 (the middle part of the second interval of one third). Overall then, at the second stage we have discarded all those numbers between 0 and 1 that have a 1 in either the first or second place in their ternary expansion.

After the nth stage we have similarly discarded all those numbers in I that have a ternary expansion that contains a 1 anywhere in the first n places. The Cantor Middle Third Set therefore comprises all those numbers in the unit interval whose ternary expansion consists entirely of 0's and 2's. For example the number $0 \cdot 202020 \cdots$ lies in C. In base ten this is the fraction $\frac{3}{4}$.* Indeed there are uncountably many members of the Cantor Set. To see this we note that there is a one-to-one correspondence between members of C and binary expansions of numbers in I, obtained by changing every instance of 2 in the expansion of a number in C into a 1. For instance under this correspondence $\frac{3}{4}$ is associated with the binary number $0.101010 \cdots$. Since there is one such binary expansion for each number in the unit interval, we arrive at the remarkable conclusion that there is a one-to-one correspondence between the Cantor Set C, and the interval I. In other words, C has just as many members as does the entire real line. From the point of view of number of points therefore, C is as large as it could possibly be, even though its measure is 0.

What is more, far from being dense, C is *nowhere dense*. Recall that by saying that a set like the rationals is dense, we mean that whenever we take a real number a, there are rationals numbers to be found in any little interval surrounding a. We say that any *neighborhood* of *a* contains members of the set or rationals. The Cantor set has quite the opposite nature—numbers not in *C* might live their lives in the real line without ever coming across any members of *C*, provided they confine their experiences to a narrow enough locality around where they live. To see this, take any number *a* that is *not* in *C*, so that *a* has a ternary expansion that contains at least one 1: $a = 0 \cdots 1 \cdots$, with the 1 in the *n*th place say. For a sufficiently tiny interval surrounding *a*, the numbers *b* in that interval have a ternary expansion that agrees with that of *a* up to places beyond the *n*th, and so will also *not* be members of the strange set *C* as their ternary expansions will also contain at least one instance of 1.

On the other hand, any member *a* of the Cantor set will not feel too isolated, for when *a* looks out into any interval *J* that surrounds it in the number line, however small, *a* will find neighbors from the set *C* living alongside it (and numbers not in *C* as well). We can specify a member *b* of *J* that lies in *C*, by taking *b* to have a ternary expansion that agrees with *a* to a very large number of places, but with no entry being a 1. Indeed there are uncountably many members of *C* in *J*. However, as we mentioned above, numbers not in *C* can live their entire lives on the number line and never set eyes upon a member of *C*, provided they are happy never to look very far into the distance.

In conclusion, the Cantor Middle Third Set *C* is as numerous as can be and, to the members of the *C* club, their brothers and sisters are to be seen all around them wherever they look. To the numbers not in *C* however, *C* hardly seems to exist at all. Not one member of *C* is to be spotted in their exclusive neighborhoods, and the set *C* itself has measure zero. To them, *C* is almost nothing.

chapter 12

Application of Number: Codes and Public Key Cryptography

Secret codes, or ciphers as they are known in the trade, are one thing guaranteed to capture the imagination of people, especially children. As a child there seems to be little that you have all your own that adults cannot gain access to and take away if they choose. Having your own way of communicating with one or two trusted friends that no-one else can understand gives little ones a rare chance to walk tall, and feel superior in a world not even parents can enter.

The most important application of codes however has been, up until very recently, in the military. Nowadays though, multifarious forms of coding are used in the electronic transfer of information. Some of this, such as the transmission of personal details, is still secret and protected but much is publicly available code the purpose of which is not secrecy, but rather the free movement of data around the world.

Examples from History

The first military code put to serious use was perhaps the so called Caesar cipher. The purpose of this cipher is simply to allow written messages to pass between commanders with some degree of security. If the messenger is captured, he himself will not divulge the content of the message, as he could not himself read it. Even if the message itself is captured, it could not be deciphered by the enemy, at least not on the battlefield. On the other hand, the proper recipient of the message needs to be able to decipher it quickly and accurately so the cipher must be readily decipherable by those in the know.

The cipher attributed to Caesar is indeed very simple for it involves shifting the letters of the alphabet along three places. A message can then be quickly deciphered, especially if one has the shifted alphabet before ones eyes:

A B C D E F G H I J K L M N O P Q R S T U V W X Y Z
D E F G H I J K L M N O P Q R S T U V W X Y Z A B C

In this Caesar cipher, the message CROSS THE RUBICON (this is known as the *plaintext* message) is enciphered as FURVV WKH UXELFRQ (called the *ciphertext* message). This might be enough to confound the enemy, at least the first time around. However it is not very secure, and indeed if the enemy knew, or guessed that the cipher was based on an alphabet shift, the code could well be cracked in a minute or two upon intercepting even a short message like this one. Indeed once the enciphered form of one single letter is correctly guessed then the whole code is blown as the cyclic shift in the alphabet is revealed: for instance if we guess that $A \rightarrow D$ when enciphered, and we know that the cipher is a simple Caesar shift, then the key to the cipher is there for all to see.*

A more difficult cipher is to swap each letter with another in no particular pattern. In this way if the enciphered form of a letter such as I or A is guessed (often an easy task as these two are the only one-letter words) we cannot immediately find the rule for the rest of the cipher because there is none. The arbitrary nature of the substitution is an inconvenience for the code users as well as it can be difficult to remember how to form the cipher. Mistakes will be made unless the secret cipher is written down and then it could easily fall into the wrong hands. A clever way around this is to replace each letter not with another letter, but another symbol drawn from a simple pattern. Those who enjoy privileged knowledge of the cipher can then reconstruct the code by re-drawing the diagram as necessary and destroying it afterwards. The pattern many of us will have seen as children is shown in Fig. 12.1.

The symbols are then drawn from the picture in a natural way (Fig. 12.1(a)).

This idea was the basis of the famous Sherlock Holmes story of *The Dancing Men* in which members of a secret fraternity threatened and coerced one another through a substitution code like this one but based on stick figures. Sinister messages appeared in all sorts of places, frightening the heroine almost out of her wits. However, it did not take Holmes very long to crack the code and turn the tables on the villains involved. He did this by using

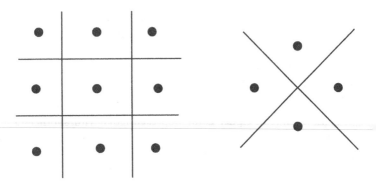

Figure 12.1. A simple pattern used to make
a substitution cipher.

frequency analysis, pattern matching, and trial and error until all
was revealed.

Given a fairly long intercepted message encoded as a simple
substitution cipher, it is not hard to spot the true meaning of
letters. The symbols for I and A are likely to occur in isolation
and common letters such as E and T will have equally common
symbols substituting for each of them. From this, short words can

$\underline{\rule{0.6cm}{0pt}}|$ = A $\underline{|\rule{0.4cm}{0pt}|}$ = B $|\underline{\rule{0.5cm}{0pt}}$ = C ... $\ulcorner$ = I $\overset{\bullet}{|}$ = J

$|\bullet|$ = K ... $\ulcorner\bullet$ = R $\vee$ = S ... $\mathbb{V}$ = W ... $\langle\bullet$ = Z

Figure 12.1a.

be guessed, giving more of the cipher and the secret is quickly blown wide open. Holmes's opponents, although they threw in some red herrings along the way, soon fell prey to this kind of analysis of their messages.

Nonetheless, by the 16th century these basic ideas had been taken further to develop military codes that were considered impregnable in their day yet could easily be deciphered by those who held their key. The main type, which stood defiant for several centuries, goes by the name of the Vigenère cipher. Its beauty is that the key is simply a single word, such as LIBERTY. Any unauthorised interceptor, *even one who knows that his enemy is using a Vigenère cipher*, will have the greatest of difficulty unravelling the code without the secret code word. Indeed it was widely accepted that cracking these codes was a practical impossibility and so was not even worth attempting directly. The only hope lay in somehow acquiring the code word. This could be any string of letters at all so the system looked completely secure to those who used it with due care and attention.[65]

This is how it works. Each letter of the key word, which is written vertically, represents the first letter in a simple Caesar cipher. We then encipher the first letter of the message using the first cipher, the second using the second, and so on, starting the cycle of Caesar ciphers over again once we reach the end of the key word. For example, suppose our plain text message is

A MAN A PLAN A CANAL PANAMA[66]

---

65 The idea seems first to have been formulated by Leon Battista Alberti of Florence in a visit to the Vatican in the 1460's.
66 A famous example of a complex palindrome—try reading it backwards!

Figure 12.2. Vignere cipher table based on
LIBERTY.

Using LIBERTY as our watch word, the sender and legitimate
receiver of the message would set up a cipher table as shown in
Fig. 12.2.

The initial A of the message is then enciphered as L; the word
MAN is enciphered using the 13th letter of the second cipher,
the first of the third, and the 14th of the fourth respectively, giving the encoded form of the word as UBR. Continuing in this
fashion, we discover the full enciphered message as shown in
Fig. 12.3.

We repeat the key word above the plaintext message as a
reminder of which of the seven shifted alphabets to use in the
encoding for each letter.

L I BE R T YLI B ERTYL I BER TY

A MAN A PLAN A CANAL PANAMA

L UBR R IJLV B GRGYW XBRRGY

Figure 12.3. Plain and enciphered text.

Immediately it is clear that the codebreaker meets some new obstacles. The standard trick of assuming that an isolated letter represents either the word A or I is still valid, but we see that the three instances of the letter A in this case are enciphered differently on each occasion, sowing the seeds of real confusion in the mind of the codebreaker. Simple frequency analysis will also be found wanting, the real frequencies being disguised by the changing nature of the code throughout the message. Is there any way of ever tackling such a perplexing cipher?

Indeed there is, and the first to show that these ciphers could be cracked was the English mathematician Charles Babbage (1791–1871). Babbage is today better known as the man who designed the first 'calculating engine'. His desire to perfom difficult astronomical calculations 'by steam' led to the design of his 'difference engines'. He commanded such respect that the government spent enough money to build two battleships in a vain attempt to make his designs a reality. However, 19th century technology was not up to the task and the project was a failure. Babbage's esteemed reputation may have in part been based on him being the foremost code breaker of the day.

Babbage's ability to deal with Vigenère ciphers came about in a rather roundabout fashion. A Bristol man named John Thwaites rediscovered a version of the Vigenère cipher and published his idea, hoping to patent it. Babbage heard of this and explained, somewhat dismissively, that this type of cipher was hundreds of years old and could be found in books on the subject. Somewhat stung by this put down, Thwaites called upon Babbage to crack the code. This challenge was born of wounded prided, as Babbage had not claimed that he could decipher Thwaites' code, just that the idea was nothing new. Babbage nonetheless warmed to the task and developed techniques that could break the code down. For some reason he did not publish his work (perhaps the British Government persuaded him not to broadcast the fact, hoping to gain a military intelligence coup at some later date). Instead the public credit went to a Prussian army officer, Friedrich Willhelm Kasiski, who independently devised the same technique and published it in 1863.

It is not too hard to see how we might go about attacking a Vigenère cipher. It is, after all, just a cycle of Caesar ciphers, which themselves succumb quite easily to frequency analysis. Indeed if we happened to know, or to guess, the *length* of the key word in the Vigenère cipher, we already have found a crack in the fortress. In our LIBERTY cipher the length of the cycle is seven, which means that an enciphered message consists of a cycle of seven Caesar ciphers. Therefore in focusing on the letters in positions $1, 8, 15, \cdots, 1 + 7k, \cdots$, we are dealing with a simple Caesar cipher. If we can identify one of the frequently occurring letters in this sequence, such as e or t, we shall soon discover that A has been shifted to L, B to M, and so on. By attacking the other embedded cycles in the same way, we could discovery the key word, LIBERTY, from which point the entire secret code would open up to us.

Of course we would not know the length of the keyword, so generally we would be in for a lot more work. This rudimentary analysis though is enough to show that a short simple word leads to a Vigenère cipher that is quite vulnerable to the cryptoanalyst. A one-letter key word corresponds to a simple Caesar cipher and a short key word would lead to too much repetition to be really secure. Certainly long conversational messages containing many common short words such as THE, AND, IT and the like would leave many clues that would be seized upon and exploited by intercepting agents.

Although inconvenient, it would not be too hard for the users of the cipher to memorize quite a long key:

CARRYTHELADTHATWASBORNTOBEKING

is an easily remembered key of length 30. Certainly the analyst would need to intercept a lot of message text before the patterns of ordinary language would be visible in a Vigenère cipher with very long key words. However, long intercepted ciphertexts do eventually leave traces of the length of the key word. For example, suppose the name *London* was used many times in an enemy plan. Although enciphered in many different ways, eventually the name *London* would be encoded in the same way more than once so that the interceptor would see duplicated enciphered text. Using our LIBERTY cipher for instance and beginning from the first letter of the key word we would encipher *London* as WWOHFG. Suppose that the interceptor spotted two instances of this strange string

WWOHFG separated by, let us say, 21 symbols from the beginning of the first string to the second. What would this represent?

It could just be a coincidence, for it may be that two completely different words were translated to the same string due to them being enciphered using different Caesar ciphers. This certainly can happen with very short strings of up to three symbols but becomes progressively unlikely with longer strings. Repetition of a six-letter string like this one would get our intercepting agent very excited. If the spy assumes what is likely, that WWOHFG represents two instances of the same word, then the separation of any two instances of this enciphered word in the ciphertext must be some *multiple* of the length of the key word. Since this separation is 21 spaces, she infers that the key word has length either 3 or 7 (the correct value) or 21. This is a real breathrough—she can now start working on the ciphertext using frequency analysis on the strings of every third, every seventh and then, if necessary, every 21st symbol. If she has got her hands on a good long sample of ciphertext, the key word should soon emerge when she looks for cycles of length seven. In this way the vulnerability of Vigenère ciphers is revealed and they are now regarded as too weak to be used in serious enciphered transmission.

Unbreakable Codes

Is it possible to devise a code so strong that it is absolutely unbreakable? The short answer is yes, (although you need to hedge this affirmative answer with one or two 'ifs'.) Indeed this can be achieved in practice by following the idea behind the Vigenère cipher to its natural conclusion. This is what Joseph Mauborgne of the US crytpographic service did around the time of the First World War.

As we have already pointed out, the weakness of the Vigenère cipher lay in the key word being short and recognizable. The answer then was to make it long and unrecognizable. But how long? Longer than any message you would ever send. To make it unrecognizable, we make the key word completely random. The result of this approach is known as the *one-time pad* cipher.

The sender and receiver each need identical copies of the one-time pad, which consists of no more than a very long totally random string of letters from the alphabet. Only they possess this super key word. The secret message is then sent in whatever way convenient using the one-time pad in the Vigenère fashion. Since the key word never ends (or more precisely does not end before the message is concluded) there is no cycle of ciphers. Since each individual letter in the key word is random, and bears no relation to any other letter, the string that is transmitted is itself a totally random string. After the message is transmitted the sender destroys the pad, as does the receiver after he has deciphered the message.

Although cumbersome, the method is secure. If the enciphered message is intercepted during transmission it is of little use to the unauthorised interceptor without access to the one-time pad. He may be able to tell something about how long the message is, but little more. Even the lengths of individual words can be masked—symbols like punctuation marks and spaces can themselves be given a symbol in an augmented alphabet. The one-time pad could then be a random string from this enhanced alphabet, completing disguising the structure of the grammar in the transmitted message.

In principle, all aspects of the message can be written in binary code—the message then becomes a string consisting of the symbols 0 and 1, which is disguised by adding to it a competely random binary string as represented by the one-time pad. If the message digit were a, and the random digit in the corresponding random string were b, then the transmitted digit would be $a + b$, where this sum is calculated according to the rules of arithmetic modulo 2: that is $0 + 1 = 1 + 0 = 1$ and $0 + 0 = 1 + 1 = 0$. For instance, if the message were simply the string of ten consecutive 1 symbols 1111111111, and the first ten digits on the one-time pad were 0111011011, then the transmitted string would be that of the random string with the digits 0 and 1 interchanged throughout: 1000100100. The unauthorised interceptor is left holding a random string that contains no information, which, in isolation, is meaningless.

Even if the eavesdropper happened to know part of the message, the intercepted string would be of no use to him in deciphering the remainder as there is no relationship whatever between the remainder of the transmitted string and the remainder of the message—the connection is a totally random substring on the one-time pad. He cannot decipher any further without getting hold of that pad.

Although completely secure, the one-time pad is used for only the highest priority intelligence, as the production of a large number of pads and the care that must go in to ensuring they are never copied and fall into the wrong hands soon becomes excessive.

A very secure cipher that can be produced without too much difficulty is a *book cipher*. This involves both parties holding copies of a very long piece of text, a book perhaps. The book is the key to the whole cipher and this must remain secret. For this reason, it would be best if the 'book' is written by the code makers themselves—no literary merit is required, indeed the more arbitrary and nonsensical the better.

The words of the book are then numbered 1, 2, ⋯ and so on up to however many words can be produced. If the sender wishes to code the message PAP, she starts reading the book and follows through till she find the first word beginning with P: it may be the 40th word, in which case the plaintext P is enciphered as the number 40. Since the next letter is A, she would find a word beginning with A, it might be 8, so that would become the next cipher symbol. To encipher the final P, she would locate the next word in the text beginning with P, it might be word number 104, and so her enciphered message would be 40 8 104. Without the 'book' this is a near impossible code to break, even if long messages are intercepted.

To be as secure as possible, the enciphering should involve always going forward in the book and, after enciphering each symbol, a good practice is to jump to the midline of the next paragraph before continuing the search for a suitable word. This ensures that there is little or no correlation between the words that are used in forming the cipher by separating them by large near-random distances in the text. Although the text itself is being used up very wastefully, words are pretty cheap. The underlying idea is similar to the one-time pad as the first letters of the words of the text are being thought of as a random string from the alphabet and the message just tells the recipient which letters to pick out of this string in order to form the plaintext message.

The most infamous example involving a book code is that of the so-called Beale Treasure, a secret message supposedly describing the location of a hoard of gold worth millions. One section of the message, a book code based on The *Declaration of Independence*, offers a tantalising story of buried loot. The rest of message is presumably coded using another book or books. It has never been deciphered! [67]

New Codes for a New World of Coding

Until the early 1970's the clandestine world of the cipher (secret code) had not fundamentally changed for thousands of years. To be sure, the codes and the code breakers had progressed in leaps and bounds. The heroic work of Alan Turing and the codebreakers at GCHQ in England in cracking the German Enigma codes is an inspiring story that still leads to furtive plots.[68] The underlying idea, and the assumptions that underpinned it, had however not altered in all that time. The purpose of a cipher was for the sender to transmit to his chosen receiver a message which, while traveling in the public domain, was vulnerable to interception. However, the transmission was of no use to the receiver unless he possessed the key to the cipher. Indeed, all ciphers had the common feature that secure messages could not be passed back and forth unless those conducting the secure conversation had, at one time, exchanged the key to the cipher in secrecy.

It was presumed that this was an implicit Principle of Coding Theory: to be effective, the key to a cipher must change hands. Around 1970 however, mathematicians began to question this and showed, with an elegant argument, that this 'principle' was not well

67 It may however be an elaborate hoax: see Simon Singh's *The Code Book*.
68 For example a British TV presenter, Jeremy Paxman, recently wound up with a stolen *Enigma* machine on his desk.

founded. The counter example is typical of mathematical argument but involves no mathematics whatsoever—just a little free thinking. Let us present the standard scenario.

The three ficticious characters involved in secret transmissions traditionally go by the names of Alice and Bob with Eve, the eavesdropper, intercepting their messages and generally causing mischief. Perhaps because of the name, Eve is usually regarded as the evil figure in the drama although this is quite unfair: Alice and Bob could be hatching plots of their own and Eve represents a benign intelligence service striving to protect citizens from the conspiratorial schemes of the other pair.

Be that as it may, transmission of a secure message from Alice to Bob does not in itself necessitate the exchange of the key to a cipher, for they can proceed as follows. Alice writes her plaintext message for Bob, and places it in a box that she secures with her own padlock. Only Alice has the key to this lock. She then posts the box to Bob, who of course cannot open it. Bob however then adds a second padlock to the box, for which he alone possesses the key. The box is then returned to Alice, who then removes her own lock, and sends the box for a second time to Bob. This time Bob may unlock the box and read Alice's message, secure in the knowledge that Eve could not have peeked at the contents during the delivery process. In this way a secret message may be securely sent on an insecure channel without Alice and Bob ever exchanging keys. (Eve still could of course simply steal the box, then neither she nor Bob would know Alice's message—this corresponds to a direct physical attack on Alice and Bob's communications medium.) This thought experiment shows that there is no law that says that a key *must* change hands in the exchange of secure messages. The padlocks could be regarded as metaphors. Alice and Bob's 'locks' might be their own coding of the message rather than a physical device separating the would-be eavesdropper from the plaintext message. This represented a fresh way of looking at an age old problem.

Simultaneous Key Creation

The story of the padlocked box sets the scene for a tantalising mathematical problem. Is it possible for Alice and Bob to set up a secure cipher between them without ever meeting one another or making use of a third party to act as a go between? After all, the practical problem that had dogged cipher applications from the beginning was that of key exchange—the initial transfer of the key to the cipher between the interested parties. In principle it was solvable: the key simply had to be exchanged with careful attention paid so that it did not fall into the wrong hands along the way. However, in practice, especially in the commercial world, thousands of people wish to talk to one another in confidence and cipher keys needed to be changed often in order to maintain the integrity of the system. In the real world the sheer effort that needed to go into secure key exchange proved to be a major cost and made widespread secure communication impossible.

Our first impulse might be to create a mathematical version of the padlocked box, the lock being a metaphor for an encryption and its key the decryption. Alice takes her plaintext message M and encrypts it, sending the message in Alice's cipher, $A(M)$ to Bob. Neither Eve nor Bob can make anything of this. Bob then puts his padlock on the box in the form of a further encryption using his own secret cipher and then send the doubly encrypted message, $B(A(M))$ back to Alice. Again Eve can make nothing of this gibberish and Alice then has the cipher form of the doubly padlocked box back in her hands. Now Alice has a problem. Applying her decryption algorithm to recover $B(M)$ from the doubly encrypted message $B(A(M))$ may not work. It depends on whether the cipher operations of Alice and Bob can be carried out in either order and yield the same net result. In general they will not.

Most mathematical operations will not *commute* in the way required. To take a very simple example, suppose that the plaintext message is the number 6 and that Alice's way of disguising her message is simple to add the number 4 while Bob's secret cipher involves doubling the number. Alice sends $6 + 4 = 10$ to Bob. Bob sends $2 \times 10 = 20$ back to Alice. If Alice now tries to remove her lock by carrying out her deciphering operation, subtracting 4, she will return the number 16 to Bob. Finally Bob tries to undo his cipher by dividing by 2 and winds up with $16/2 = 8$. But this is wrong—he was supposed to end up with the plaintext message of 6. The trouble is the two ciphers, that is the two mathematical padlocks, have interfered with one another's operation.

This seems to be only a technical hitch. Surely we can get around this by finding ciphers that can easily glide past one another. For instance, both Alice and Bob could encipher their message by adding on their own personal secret number (which could be huge). If for instance Bob added 2 instead of multiplying by 2 the problem vanishes: Alice would take her message (the number $M = 6$), send it disguised as $6 + 4 = 10$, Bob would return $10 + 2 = 12$ to Alice, who would then subtract her secret number and reply with, $12 - 4 = 8$, and finally Bob would subtract his secret number to reveal the original message $8 - 2 = 6$.

However, we must not forget Eve. Put yourself in her place. Eve intercepts all these numbers and knows, or at least suspects, that the cipher of both Alice and Bob involves addition of a secret number. She intercepts the first communication, Alice sending the number 10 to Bob. Next she intercepts Bob's reply, the number 12 and immediately she cracks Bob's cipher for it is the number $12 - 10 = 2$. Next Eve observes that Alice has converted Bob's message of 12 to 8, showing that her secret cipher number is $12 - 8 = 4$. Having cracked both ciphers Eve now has no trouble deducing that the plaintext message of Alice must have been $10 - 4 = 6$. What is more it would not help Alice or Bob to replace their secret cipher numbers with huge ones for Eve could still use the same method to reveal their values. Simple addition is too simple minded a basis for a cipher to defeat the resourceful Eve.

In the mid 1970's Whitfield Diffie and Martin Hellman took a different slant on the idea of a mathematical copy of the double padlocks for secure key exchange. If only, they mused, it were possible for Alice and Bob to cast a spell that would magic up a key—the same key—in the security of their own homes. They could then use it to converse, safe in the knowledge that the nefarious Eve could not listen in.

Again a key can always be coded in terms of numbers, indeed a single number will suffice, provided it is big enough. Therefore their search was for a way for Alice and Bob to communicate just enough information for them to create the key number in their secure environments. The approach involved a process that was assumed to lie in the public domain. However, each of Alice and Bob have their own secret ingredient that is never revealed to anyone at all, not even one another. Somehow they must exchange just enough information to cook up the same cipher key, which will then be the basis of further secure communication. Eve will know Alice and Bob's methods and eavesdrop on all their insecure dialogue yet, despite having massive intellectual resources and computing power at her disposal, she will not be able to reproduce the key to Alice and Bob's communications. (Put in this light, we can understand why governments the world over are not keen on just anyone having access to such good ciphers.)

The Diffie-Hellman approach is conceptually simpler than the doubly padlocked box as it involves enciphering but no deciphering to create the key – locking but no unlocking, making the process only half as complicated. Impossible, we may think, but what may sound far fetched can be made more plausible by means of another simple metaphorical example.

As their secret key, Alice and Bob are going to manufacture an exact color shade of paint.[69] Each takes one litre of white paint and mixes it with another litre of paint of a color that only they know: Alice might use her own secret shade of scarlet, Bob his own peculiar blue. They then arrange a rendevous to exchange paint cans: Alice handing Bob two litres of pink paint, Bob giving Alice a two-litre pot of pale blue. They may even taunt their relentless adversary Eve by inviting her to their tryst and giving her an exact replica of each of the two-litre cans of colored paint. Alice and Bob then return to their own homes. Alice takes Bob's can and mixes with it one litre of her special scarlet paint. At the other end, Bob mixes in a litre of his blue into the can that Alice gave to Bob. Both Alice and Bob now have three-litre mixtures of a particular shade of purple, consisting of one litre of each of white, scarlet, and blue, and it is this exact shade that is the secret key to their cipher.

69 This enlightening example is from Simon Singh's *The Code Book*.

Eve on the other hand is left holding the cans and is stymied. She cannot unmix the paint to find out the exact shades of scarlet and of blue that Alice and Bob have used. Even more frustrating, even though she has the two-litre mixtures of red & white, and of blue & white, it is not possible for her to create from them a paint mixture in which the ratios of white to red to blue are 1 : 1 : 1, which is what she wants to do in order to create the exact shade of purple she needs that represents Alice and Bob's key. (This is because whatever mixture she concoctes from the two cans will always be half white.) Importantly this was all done without any deciphering on the part of Alice and Bob (they didn't need to unmix paint). Indeed the common key they have created did not even exist until after each had returned to their own secure environment to conjure it up. If only Alice and Bob could talk with paint, then the key exchange problem would truly be solved!

Diffie and Hellman had a neat idea but the challenge was to produce a mathematical version of the paint mixing exchange. Crucially, the operations involved must commute with one another: when mixing paint, the final outcome depends only on the ratio of the colors we use and not on the order in which the paints are mixed together. The enciphering processes must likewise be able to slip past one another to produce the same overall effect.

One method that might occur to Alice and Bob would be to base their secret cipher on a power of 2 (not necessarily integral). For example, Alice might select as her secret number $a = 1.71$ while Bob chooses $b = 2.92$. Alice then sends to Bob (and presumably Eve) $2^a = 3.2716082$, while Bob sends Alice, $2^b = 7.5684612$. Alice and Bob then create the secret cipher based on the number 2^{ab}. In Alice's case she takes the number Bob sent her and raises it to the power a to find that $(2^b)^a = 2^{ba} = 31.849526$. Bob likewise creates the same number by taking Alice's given number 2^a, and raising it to the power b to get $(2^a)^b = 2^{ab} = 31.849526$. Since the operations of exponentiating to one power and then another do commute, Alice and Bob have created the same key to their cipher code.

But what of Eve? She has intercepted the values of both 2^a and 2^b and needs to find the value of 2^{ab} to be able to decipher Alice and Bob's future conversations. Unfortunately for Alice and Bob, if Eve is any sort of mathematician, she will be able to find the values of both a and b and then the required 2^{ab} with ease.[70] Nonetheless, the idea of repeated exponentiation was successfully used by Diffe and Hellman to allow Alice and Bob to use a method akin to this to create a mutual key that any outsider could recreate only with the utmost difficulty. Their method exploited the added ingredient of modular arithmetic.

Once again Alice and Bob choose a base number, for the purposes of the example we take it to be 2, and once again Alice and Bob choose one number each known only to them personally. This time we even insist that they select ordinary positive integers: let us say Alice chooses $a = 7$ and Bob goes for $b = 9$. However there is now to be an extra ingredient, another number p, which is also assumed to lie in the public domain: let us suppose that $p = 47$. Alice now computes 2^a as before but this time the number she transmits is the *remainder* when this number is divided by p. In this case she finds $2^7 = 128 = 2 \times 47 + 34$, so the number 34 is sent over an insecure channel to Bob. Similarly Bob computes $2^b = 2^9 = 512 = 10 \times 47 + 42$, and transmits 42 to Alice.

What Alice now does in the security of her own home is calculate the remainder when 42^a is divided by p, while Bob calculates

70 Writing c for *known* number 2^a, Eve then has $a = \log_2 c$.

the remainder when p is divided into 34^b. Alice and Bob will both end up with the same number, the same key, as in each case the net result will be the remainder when 2^{ab} is divided by p.* Alice will find that the remainder when 42^7 is divided by 47 is 37, and so will Bob when he divides 34^9 by 47. Alice and Bob have now created a shared key, the number 37.

Eve on the other hand is left frustrated. Her mathematical problem is this; she does not know the values of a or b but she does know that 2^a and 2^b leave respective remainders of 42 and 34 when divided by 47. The key is to find the remainder when 2^{ab} is divided by 47. This is much more difficult than her previous problem that involved no arithmetic of remainders. In the original attempt where Alice and Bob exchanged powers of 2, Eve would have little difficulty homing in on the actual values of a and b. Given that $2^a = 3.2716082$ we see immediately that a must be between 1 and 2 and Eve can play the higher-then-lower game to approximate the value of a better and better. She would test the values $a = 1.5, 1.6, 1.7, 1.8$ and discover that $2^{1.7} < 2^a < 2^{1.8}$, telling Eve that $a = 1.7\ldots$. Then she would continue the hunt in the second decimal place and soon discover that Alice used $a = 1.71$. In the same way, Eve would soon know Bob's secret number was $b = 2.92$ and she would be away.

However, by contrast, the remainder when higher and higher powers of a are divided by a fixed number p behaves much more erratically, rendering this approach useless. In reality there is not much alternative to testing all the possible keys and this Eve can try: she can compute $2^1, 2^2, \cdots$ and find the remainder when each is divided by 47 until she hits on a value that matches the remainder when Alice's 2^a is divided by $p = 47$. Then she could calculate the value of the key in the same way that Alice did and Eve will have breached the security of Alice and Bob. In our little example, this approach is clearly possible but in practice, Alice and Bob can use numbers so large that this approach becomes infeasible. Roughly speaking, unless Eve has access to much, much stronger computational power than Alice and Bob, Eve will not be able to break into the key for a very, very long time. She will have to give up and try another approach.

And there are other evil things for Eve to contemplate. In her frustration she may try to mislead Alice and Bob by sending messages of her own purporting to come from them. Alice and Bob still need to be on their guard.

Opening the Trapdoor: Public Key Encryption

The Diffie-Hellman key exchange was an exciting development but a fresh ideas was still needed, the reason being that the manner in which security codes are used, for example on the internet, is very different from the traditional use, something that might not be clear at first glance.

For example, when a customer entrusts their personal details to an internet provider, address, phone, credit card number and so forth, they need to be sure that this information will not be intercepted and transferred elsewhere. The safe transfer is effected through the sensitive information being enciphered. However, the customer knows nothing of this cipher so how is this done? It comes as no surprise to learn that this is carried out automatically on the customer's behalf—the buyer need have no knowledge of the code being used and may not be even be aware of its existence. There is potentially a big problem with this. The encoding has to be done *before* transmission, otherwise there is no point and no security. This means that the enciphering process lies in the public domain. It may not be readily visible to the consumer, but it is present in the system to which the general public have access, so it cannot be regarded as secure. If an unscrupulous party gains access to the enciphered transmissions, and also knows how to encipher the message, surely it will not be too hard to reverse the process and decipher the original message? This would be disastrous and make all such transactions insecure, rendering confidential internet traffic an impossibility.

For example, if the enciphering process was a Vigenère cipher of some kind, perhaps even a one-time pad, and the enciphering pad was accessible then the interceptor could decipher the message just as easily as the proper receiver. Surely once Eve knows how to encipher messages, she will be able to decipher them as well, and undermine the system? This would certainly be the case with all the codes that we have introduced to this point. The problem calls for a new way of doing things. What is required is to devise a code for Alice, which she can place in the public domain so that anyone can use it to send her messages but, somehow, she is still the only one who can decipher the coded message—the 'public' key is one that can lock, but not unlock the vessel containing her secret. No so called Public Key Cryptosystem is possible until a solution to this problem is found. A solution however held the promise of a new era in safe transfer of private information.

In the 1970's a number of people hit on this idea and realized its potential importance. However, to bring the idea to fruition involved the invention of a *trapdoor function*. Each user would need such a function f that would be in principle available to everyone who could then calculate its values $f(x)$. However, the owner of the function, Alice, would know something vital about it that allowed her to decipher and recover x from the value of $f(x)$. What is more, other people, even though they knew how to calculate $f(x)$, must not be able to deduce this key piece of information however hard they try. This seemed a tall order.

Nonetheless, it was achieved by Clifford Cocks soon after joining the British Intelligence organization GCHQ in Cheltenham in 1973. After being introduced to the idea of public key cryptography by his colleagues he invented a suitable system in about an hour. He used his knowledge of Number Theory to devise a suitable trapdoor function with the required one-way property: given x, anyone could calculate $f(x)$ but given $f(x)$, it was near impossible to recover the number x unless you were in on the secret of its structure.

The mathematics that Cocks exploited was not very deep and will be explained below. It was however absolutely pure mathematics and, it seems, no-one but a pure mathematician would ever have come up with it. His method is the basis of today's public key cryptography.[71]

Unfortunately, Cocks worked for a secretive government organization so his great breakthrough was never released into the public domain. Instead, the same ideas were stumbled on and exploited by a collection of half a dozen mathematicians and computer scientists working in the USA a few years later. The names usually associated with the discovery and development of public key

---

71 Secure key exchange and public key crytpography are closely related ideas and indeed in Britain were discovered in the opposite order to the USA: at GCHQ Malcolm Williamson discovered the idea of Diffie-Hellman key exchange contemparaneously with the American pair while trying to find a flaw in Clifford Cocks' creation of a public key cryptosystem.

crytpography are Diffie, Hellman and Merkle along with Rivest, Shamir and Adleman from whose initials the name RSA codes derives.

As we have said, the idea of a trapdoor function is the key to it all but having the idea is not enough. Those who became enmeshed in the search for a suitable trapdoor cast around wildly, devising all forms of fantastical procedures in the search for this their Holy Grail. However, by far the strongest candidate that has been devised so far, and the one on which nearly all commercial encryption is currently based, is that of Clifford Cocks and rests upon the observation that it is exceedingly difficult in practice to find the prime factors of a very large number even though, in principle, the problem is simple to solve.

The principal ingredient of Alice's RSA private key is a very large pair of prime numbers, p and q. (In real life these numbers are up to 200 digits in length.) In order to use Alice's public key however, Bob does not need p and q but rather the product, n of these two primes: $pq = n$. This represents the first step in the process. The next key step however is to invent a trapdoor function $f(x)$ that can be calculated as long as we possess n but has the property that, give the number $f(x)$, it is a practical impossibility to recover x without the two magic numbers p and q.

Practical experience had shown that recovering p and q from n took a prohibitive amount of computing power. However, taking the next step, finding a suitable function $f(x)$, required both diabolical cunning and familiarity with the theory of numbers.

Before proceeding it is worth taking a moment to contemplate how revolutionary all this is as it completely contradicted the received wisdom as to what constituted applicable mathematics. Pure number theory was a field regarded by everyone as being among the most useless areas of mathematics—indeed G.H. Hardy took positive delight in this notoriety. Properties of integers could be significant in applications but the theory involved was always quite rudimentary, almost amounting to mathematical common sense. The mathematics that Cocks and the others used however is based on the Euler totient function (see Note 48) which, although centuries old, is by no means trivial and indeed forms the basis of the subject. Today the RSA program is the most used piece of software on Earth and it is squarely based on the ideas of Euclid, Fermat and Euler and the arguments of Cocks. Mathematical ideas are often centuries ahead of their own era but when their time arrives, their impact can be revolutionary.

Alice and Bob Vanquish Eve with Modular Arithmetic

We now give an account as to how Clifford Cocks proceeded. As has been explained before, since any message can be translated into a string of numbers, the problem comes down to how Bob may securely send a particular number, let us call it M for message, to Alice without Eve finding out its value. As mentioned above, Alice's private key is based on two prime numbers, p and q that only she knows. In this toy example, which is quite representative of the real situation, we shall use the small primes $p = 23$ and $q = 47$. The publicly known product of these two numbers is $n = 23 \times 47 = 1081$. (In practice of course, p and q are huge and in any case all this is happening behind the scenes and is done invisibly on behalf of any real life Bob and Alice.)

The approach is to mask the value of M using modular arithmetic, that is to say clock arithmetic in this case based on a clock whose face is numbered by $0, 1, 2, \cdots, n - 1$. What Alice leaves in the public domain is the number n and also another number, e for encoding messages meant for her. What Bob sends to Alice is not of course M itself (for if he did then Eve would be liable to overhear) but rather *the remainder when M^e is divided by n.* For example, if Bob's message was $M = 77$ and if the encoding number that Alice tells people to use is $e = 15$, then Bob, or rather his computer, would calculate the remainder when 77^{15} was divided by $n = 1081$. This remainder turns out to be 646. (If you set out to check this with your own calculator you will find that the poor thing will complain bitterly over the size of the numbers involved. However, there are simple tricks* that allow us to calculate the *remainder* in this sum without needing to calculate the very big number 77^{15}.)

And so Bob sends to Alice his disguised message in the form of the enciphered message 646. Eve will presumably intercept this message and know that Bob's message is encoded as 646 when using Alice's public key which she knows as well as anyone consists of $n = 1081$ and $e = 15$. But how can the original message be teased back out?

For Alice, who knows that $1081 = 23 \times 47$, this is quite straightforward. For, once in possession of the prime factors of n, it is possible to determine a decoding *number d* which is found using the values of p, q and e. It turns out in this case that a suitable value of for the decoding number is $d = 135$. Alice's computer then works out the remainder when 646^{135} is divided by $n = 1081$, and the underlying mathematics ensures that the answer will be the original message $M = 77$.

Further mathematical explanation is to be found in the notes of the final chapter. I will mention however some of the mathematical niceties that need to be respected in RSA. Although it did not appear in the above description, a key ingredient in the method is the value of the number $(p-1)(q-1)$, which is denoted by $\phi(n)$, and in this case we see that $\phi(1081) = 22 \times 46 = 1012$. The encoding number e that Alice chooses in her public key cannot be competely arbitrary but must have no factor in common with $\phi(n)$. The prime factors of 1012 are seen to be 2, 11 and 23 so that e must not be a multiple of any of these three primes. This is only a very mild restriction and Alice's particular choice of $e = 15 = 3 \times 5$ is perfectly all right. The decoding number d is chosen, and this is always possible, so that the product ed leaves a remainder of 1 when divided by $(p-1)(q-1)$.* The message number M itself needs to be less than n but in practice this is no restriction as the size of n in real applications is so monstrous it can accommodate all the values of M enough to cover any real message we would ever wish to send.

To see all this in action we may illustrate with an example featuring even smaller numbers that the one above. For instance let us take $p = 3$ and $q = 11$ so that $n = pq = 33$ and $\phi(n) = (p-1)(q-1) = 2 \times 10 = 20$. Alice then publishes $n = 33$ and suppose she sets $e = 7$, which is permissable, as 7 has no factor in common with 20. The number d then has to be chosen so that $ed = 7d$ leaves a remainder of 1 when divided by 20. By inspection we see a solution is $d = 3$, for then $7d = 21$.

Now Alice has her little RSA cipher all set up. If Bob wants to send the message $M = 6$, then he computes $M^e = 6^7 = 279,936$, divides this number by 33 to find that the remainder is 30, and so Bob would send the number 30 over an open channel. Alice would receive Bob's 30 and decipher its real meaning by calculating $30^3 =$ 27, 000. Division by $n = 33$ then gives her 27, 000 $= 33 \times 818 + 6$. Again it is only the remainder 6 that is of interest as that is Bob's plaintext message.*

For the time being, RSA encryption is effective and safe but there are still ways in which Eve may try to sow seeds of confusion and that must be guarded against. It is true that Bob may now send messages to Alice safe in the knowledge that only she can understand them. But how is Alice to know that the message really comes from Bob and not some imposter? Eve, (who we always assume is hideously intelligent and does nothing all day except hatch plots to make life a misery for Alice and Bob) can easily send messages of her own to both Alice or Bob, claiming that they come from the other.

However, Bob can authenticate his messages to Alice using his own private key and Alice should not trust any message purporting to come from Bob unless it contains this so-called *digital signature*. The way Bob proceeds is as follows. He writes his personal message to Alice in plaintext in his own home. He then takes some personal form of identification, let's call it I, which could be his name perhaps together with some other personal details, and treats it as if it were an incoming message—that is to say he *decrypts* I, using his own private key, to form a string of gibberish we shall call $B^{-1}(I)$. The notation here is meant to convey the idea that Bob is *inverting* the normal procedure in that he is 'deciphering' the string I with his own private key instead of enciphering it with a public key. This is not secure, on the contrary, anyone who suspects that $B^{-1}(I)$ comes from Bob can verify this by using Bob's public key, and this is the whole point. When Alice finally receives Bob's message she will take this meaningless looking string and feed it into Bob's public key B to retrieve $B(B^{-1}(I)) = I$ again. Alice will then know the message truly came from Bob, as only he has the power to create the string $B^{-1}(I)$.

In full, Bob's computer executes the following tasks on his behalf. It takes Bob's plaintext message, M, along with his digital signature, $B^{-1}(I)$, and encrypts it using Alice's public key. The encrypted message is then sent to Alice who is the only one who can decrypt it to recover M and $B^{-1}(I)$. Finally Alice's machine will recover I using Bob's public key, which tells her that the origin of the incoming message really is Bob and no-one else.

Eve is left impotent with rage. She certainly cannot get into the message sent by Bob as she lacks Alice's private key, so she will not even be able to see the digital signature $B^{-1}(I)$ that Bob has used as authentification. She can send messages to Alice using Alice's public key, but if Alice's computer system is vigilant it will reject them as they will lack the authentification of Bob or any of Alice's confidantes. Eve cannot interfere with the communications between Alice and Bob, nor can she even talk to either of them herself. Eve is firmly locked out of Alice and Bob's world.

It seems that the pythagorean dictum that 'All is Number' reigns supreme in the world of secure communications. But is this a temporary state of affairs? There are two reasons to suspect that may be the case. First there is the general observation that the see-saw battle between the codemakers and breakers has a long history whereby the cipher makers for a time seem invulnerable, only to have the tables turned in dramatic fashion by the code breakers. We really should be prepared for this repetition in the cycle as the implicit conflict in the problem lends to it, that conflict being that the legitimate receiver needs to be able to decipher with ease what the unauthorised interceptor finds impossible to make head or tail of. On the other hand, RSA seems safe. To be sure, Eve may, and probably soon will, increase her computing capacity many times over, allowing her to crack current private keys in quick order. However, Alice and Bob will not be standing still and, just by finding ever larger primes (after all, Euclid showed us they never run out) will be able to keep Eve at bay with relative ease.

At a more practical level however, even the casual observer cannot help noticing that in wholeheartedly embracing RSA, wonderful as it is, we are putting all our commercial coding eggs in one basket. What would happen if their was a mathematical, as opposed to computational breakthrough, which made the system vulnerable? All secrecy on the internet would vanish overnight and the outcome would be economically catastrophic!

Some may appeal to the authority of the code makers and say that we have been assured that the problem of prime factorization is intractable and so its security is a mathematical fact. This however is not the case—no such absolute assurance is offered. Indeed, to the contrary, in 2002 a warning was sounded when three Indian mathematicians (Agrawal, Kayal and Saxena) rather shocked the mathematical community by proving that the problem of *primality*, determining whether or not a give number was prime, could be solved in what is known as polynomial time. Their proof in itself does not seem to pose an immediate threat to RSA but does demonstrate that the problem Eve faces in cracking an RSA cipher is perhaps not intractable in the absolute sense that does apply to some other combinatorial problems. It would be very worthwhile to find other trapdoor functions, or other methods of key exchange, which did not depend on this one mathematical trick, just in case the current invulnerablity of RSA proves to be an illusion. RSA is however not the only Public Key system in general use: for example the US National Security Agency employs ciphers based on so-called elliptic curves.

At first sight, cryptanalysis may seem unworthy of so noble a subject as mathematics for the whole desire for secret ciphers stems only from our own devious nature and not from the natural world. That judgement however is superficial. The way information is stored and transferred while being hidden from view is intrinsically interesting and has counterparts in nature in the way DNA carries with it the description of our make up. Indeed much of science involves teasing out hidden information from mere traces of what has been left behind. Mathematics in general and numbers in particular are often the link between what we can see and that which we seek to discover.

chapter 13

For Connoisseurs

This final chapter is included to highlight a little of the mathematical detail surrounding some of the claims made in the body of the text. The level of difficulty and knowledge required to understand what is on offer varies and most readers should be able to gain something from sampling from this chapter. Unlike the rest of the book however, I do make free use of mathematical symbology and the occasional passage assumes familiarity with one aspect of mathematics or another.

Chapter 1

Note 1 Page 9 *The Riemann Hypothesis and Prime Numbers*

This first note is about the deepest mathematics in the book and so therefore is the explanation that accompanies it. The *Riemann Zeta function* is a function $\zeta(z)$ of a complex variable z. It is defined for all values of z with real part greater than 1 by the formula:

$$\zeta(z) = \sum_{n=1}^{\infty} \frac{1}{n^z}.$$

The values of the zeta function on the even integers have long been known: $\zeta(2) = \frac{\pi^2}{6}$, and the outputs for other even integers are rational multiples of powers of π involving what are known as *Bernoulli Numbers*. For odd integers the values remain largely a mystery to the extent the achievement of Apery in showing that $\zeta(3)$ is irrational amazed the mathematical community, so difficult was the result. The zeta function has an intimate relation with the *Gamma function*, which is a function defined by an integral that generalizes the factorial function to other real values. This relation can then be used to extend the definition of $\zeta(z)$ for all values of z except $z = 1$. This relationship also reveals that $\zeta(-2n) = 0$ for all positive integers n. These zeros of ζ are known as its *trivial zeros*. The Riemann Hypothesis is that all the other zeros of ζ have real part $\frac{1}{2}$. It is known that the non-trivial zeros all lie in the *critical strip* of numbers with real part strictly between 0 and 1, but so far all zeros discovered actually lie on the *critical line* where $Re(z) = \frac{1}{2}$ (and it is known that infinitely many zeros lie on this line). If this 150-year old conjecture were resolved in the affirmative, then the distribution of the prime numbers would to a large extent be known. Most mathematicians have always regarded The Riemann Conjecture as the greatest unsolved problem in mathematics, even more outstanding than Fermat's Last Theorem (finally proved by Wiles in 1995).

The connection between ζ and the primes is by way of a stunning little argument due to Leonhard Euler. Let $p_1, p_2, \ldots$ be the (infinite) list of all primes, let z be any complex number (including reals of course) outside the unit circle, and consider the infinite product:

$$\left(1 + \frac{1}{p_1^z} + \cdots + \frac{1}{p_1^{kz}} + \cdots\right)\left(1 + \frac{1}{p_2^z} + \cdots + \frac{1}{p_2^{kz}} \cdots\right)\cdots$$
$$\left(1 + \frac{1}{p_n^z} + \cdots + \frac{1}{p_n^{kz}} + \cdots\right)\cdots$$

The sum of a typical geometric series in the brackets is $\frac{1}{1 - \frac{1}{p_n^z}} = \frac{p_n^z}{p_n^z - 1}$. On the other hand, multiplying the brackets out, a typical term has the form $\frac{1}{n^z}$. The value of n will depend on which prime powers are chosen (only finitely many factors are not 1, for all other infinite products are 0). Since every number is a product of primes, every possible term $\frac{1}{n^z}$ will arise. Crucially, since the prime factorization of a number n is unique, each term $\frac{1}{n^z}$ will emerge once only in the corresponding infinite sum, which is therefore $\zeta(z)$. This gives the Euler identity that eternally ties the prime numbers and the zeta function together:

$$\zeta(z) = \Pi_{n=1}^{\infty} \frac{p_n^z}{p_n^z - 1},$$

the product being taken over all prime numbers.

Simple ratios involving the zeta function also relate it to the number theoretic functions $s(n)$ (see Note 2) and $\phi(n)$ (see Notes 48 and 56).

Note 2 Page 13 *Counting and summing factors*

Let n be an integer with prime decomposition $n = p_1^{r_1} p_2^{r_2} \cdots p_k^{r_k}$, and let us write $d(n)$ and $s(n)$ for the number of divisors of

n, and the sum of the factors of n, respectively. The value of these functions is given by

$$d(n) = (r_1 + 1)(r_2 + 1) \cdots (r_k + 1) \text{ and}$$
$$s(n) = \frac{p_1^{r_1+1} - 1}{p_1 - 1} \cdot \frac{p_2^{r_2+1} - 1}{p_2 - 1} \cdots \frac{p_k^{r_k+1} - 1}{p_k - 1}.$$

Each of these formulas is simple to prove when $k = 1$. The result then follows in each case from the (unobvious) fact that d and s are *multiplicative:* a function f on the positive integers is called multiplicative if $f(mn) = f(m)f(n)$, whenever m and n are relatively prime, that is to say have a hcf of 1.

Note 3 Page 14 *The infinity of primes*

There are many proofs of this but none better than the original euclidean argument. Let $p_1, p_2, \cdots, p_k$ be the list of the first k primes, and consider the number $n = p_1 p_2 \cdots p_k + 1$. Either n is a prime, or is divisible by a prime smaller than itself, which cannot be any of $p_1, p_2, \cdots, p_k$, as if p is any one of these primes then $\frac{n}{p}$ leaves a fractional remainder of $\frac{1}{p}$. It follows that there must be a new prime, q say, that is greater that all the primes $p_1, p_2, \cdots, p_k$ and no more than n itself. In particular there can be no finite list of primes that contains every prime, and so the sequence of primes does not end. Arguments along these lines can also show that there are infinitely many primes of the forms $4n + 3$, $6n + 5$, and $8n + 5$. For the second, see Note 17 below.

The largest known prime given in the text is a *Mersenne prime*, a prime of the form $2^p - 1$, where p is itself a prime. Euclid proved that for every such number we can find an even perfect number, that being, $2^{p-1}(2^p - 1)$, and in the 18th century Euler proved that every even perfect number is of this type, so that Euclid and

Euler together established a one-to-one correspondence between the Mersenne primes and the even perfect numbers. However, we do not know whether there are any odd perfect numbers, nor do we know if the sequence of Mersenne primes runs out or not. (You would guess not, but how to prove it?) The Mersenne numbers are natural prime candidates as it can be shown that any divisor of a Mersenne number (for not all are prime, for example, take $p = 11$) has the form $2kp + 1$. This again tells us that there must be infinitely many primes, for it shows that the smallest prime divisor of $2^p - 1$ exceeds p, and so p cannot be the largest prime. Since this applies to every prime p, we conclude that there is no largest prime and the prime sequence runs on for ever.

A similar argument yields the same result through consideration of the *Fermat Numbers,* the numbers of the form $2^{2^n} + 1$ for $n = 0, 1, 2, \cdots$. It is not hard to verify from the definition that the product of the first $n - 1$ Fermat numbers is 2 less than the *nth* Fermat number, from which we infer that any two Fermat numbers are relatively prime, that is to say, have a highest common factor of 1, as any common factor of F_k and F_n must divide 2, and since the Fermat numbers are odd, that divisor could only be 1. From this we infer that there must be infinitely many different prime factors of the Fermat numbers, and so there are infinitely many primes.

The Fermat numbers are important as in the 19th century Gauss showed that a regular polygon is constructible with euclidean tools if and only if the number of sides is a product of different *prime* Fermat numbers multiplied by some power of 2. For example, since $F_2 = 17$, the regular 17-gon can be constructed using straightedge and compasses, something that Euclid never knew. However although the Fermat numbers up to $n = 4$ are prime, there seem to be no more after that. (See more in Note 18.)

Chapter 3

Note 4 Page 38 and Page 46 *Casting out nines*

The claim is that any number is equal, modulo 9, to the sum of its digits. The crucial observation is that $10^n - 1 = 999 \cdots 999$ (with n 9's), which is clearly a multiple of 9. In modular notation we write $10^n - 1 \equiv 0 \pmod 9$, and so $10^n \equiv 1 \pmod 9$. The upshot of this is that, when working in multiples of 9, we may replace any power of 10 by the number 1. (We are implicitly using the fact that an equation involving modulo equality can be treated as an ordinary equality in that we may add, subtract, multiply and raise to powers on both sides and the $\equiv$ sign is respected—you do however have to take more care when it comes to division!) It now follows that any number $a = d_k d_{k-1} \cdots d_0$ (each d_i a digit) is equal mod 9 to the sum of its digits for then

$$a = d_0 + 10d_1 + 100d_2 + \cdots + 10^k d_k \equiv d_0 + d_1 + \cdots + d_k \pmod 9.$$

At the same time we have found the justification for our divisibility test for 9: a number a is equal mod 9 to the sum of its digits, s, so that in particular, 9 is a factor of a if and only if 9 is a factor of s. Hence we may test for divisibility of a by 9 by testing the smaller number s instead. Indeed the previous argument shows that the test will work for any number m, provided that $10^n - 1$ is a multiple of m. Since this is evidently true for $m = 3$, at a stroke we have also justified the divisiblity test for 3.

Note 5 Page 48 *Divisibility Tests for* 7, 11 *and* 13

The key observation is that $10 \equiv -1 \pmod{11}$ and so $10^n \equiv (-1)^n \pmod{11}$. Then we argue similarly to the modulo 9 case:

$$a = d_0 + 10d_1 + 100d_2 + \cdots + 10^k d_k$$
$$\equiv d_0 - d_1 + d_2 - \cdots + (-1)^n d_n \pmod{11}.$$

It follows therefore that a is divisible by 11 exactly when the same is true of its alternating sum of digits.

The block-of-three test for 7 is similarly based on the observation that $1000 \equiv -1 \pmod 7$. Indeed the test will work for any number m such that $1000 \equiv -1 \pmod m$. This also includes 11 and 13, as is easily checked. As a bonus, we can also verify that $1000 \equiv 1 \pmod{37}$ (as $999 = 27 \times 37$), and so we get a similar test for divisibility by 37, except that it is simpler in that the alternating sign vanishes. For example 105, 191 is divisible by 37 by virtue of the fact that $105 + 191 = 296 = 8 \times 37$.

A related number pattern that often surprises is that when any number of the form abc, abc is divided by 7, 11, and 13 the result is always abc. This applies for example to $749, 749$ or even to $94, 094$ (by taking a to be 0). The reason for this becomes clear when we look at what is happening in reverse: $7 \times 11 \times 13 = 1001$, and the effect of multiplying the three digit number abc by 1001 is

$$1001 \times abc = 1000 \times abc + 1 \times abc =$$
$$abc, 000 + abc = abc, abc.$$

Note 6 Page 50 *Magic Constants*

The nth triangular number is $t_n = \frac{1}{2}n(n+1) = 1 + 2 + \cdots + n$. A normal $n \times n$ magic square features all numbers from 1 up to n^2. Hence each line sums to

$$\frac{1}{n}t_{n^2} = \frac{1}{n} \cdot \frac{1}{2}n^2(n^2+1) = \frac{1}{2}n(n^2+1).$$

The first five magic constants are therefore 1, 5, 15, 34, and 65 although there is no normal magic square for $n = 2$.

Note 7 Page 52 *Complementary Magic Square*

Subtracting each of the numbers $1, 2, \cdots, n^2$ from $n^2 + 1$ in turn gives the same set of numbers, $n^2, n^2 - 1, \cdots, 1$ in reverse order. Therefore the complementary square does have each number from 1 to n^2 again appearing exactly once. Take any line in the original magic square, $a_1, a_2, \cdots, a_n$. The sum of the a_i is the magic number $\frac{1}{2}n(n^2 + 1)$, by the previous note. Therefore the sum of the corresponding line in the complementary square is

$$(n^2 + 1 - a_1) + (n^2 + 1 - a_2) + \cdots + (n^2 + 1 - a_n) = n(n^2 + 1)$$
$$-(a_1 + a_2 + \cdots + a_n) = n(n^2 + 1) - \frac{1}{2}n(n^2 + 1) = \frac{1}{2}n(n^2 + 1)$$

which is again the nth magic number, so the complementary square is also a normal magic square.

Note 8 Page 54 *Picking numbers from each row and column*

The process described has you picking the numbers in such a way that each number chosen comes from a row and from a column that has not previously been sampled. At each step we cross out a new row and column so that we pick a set S, of n numbers in all, which therefore must feature exactly one from each row and column. We show that the sum of the members of S must be the magic number of Note 6.

The first members of each row form the sequence

$$1, n + 1, 2n + 1, \cdots (n - 1)n + 1.$$

It follows that the set S consists of numbers of the form: $(rn + 1) + k$, $(0 \leq r, k \leq n - 1)$. Since we choose exactly one number from each row and each column, all possible values of r and of k occur exactly once. It follows that the sum of the members of S, is given by

$$\sum_{r=0}^{n-1}(rn+1) + \sum_{k=0}^{n-1}k = n \cdot \frac{1}{2}n(n-1) + n + \frac{1}{2}n(n-1)$$

$$= \frac{1}{2}n((n^2-n) + 2 + (n-1)) = \frac{1}{2}n(n^2+1),$$

which is the nth magic number.

Chapter 4

Note 9 Page 62 *Amicable Numbers*

The 9th century Persian mathematician Thābit recorded a remarkable fact that allows us to find amicable pairs, somewhat akin to Euclid's formula for generating even perfect numbers from Mersenne primes. For $n \geq 2$, *if* the three numbers $p = 3 \cdot 2^n - 1$, $q = 3 \cdot 2^{n-1} - 1$ and $r = 9 \cdot 2^{2n-1} - 1$ are prime, then $2^n pq$ and $2^n r$ form an amicable pair. For example, for $n = 2$ we get the smallest pair of $220 = 4 \times 5 \times 11$ and $284 = 4 \times 71$.

Note 10 Page 65 *Rows of the Arithmetic Triangle*

If we write $C(n, r)$ for the number of ways of choosing a set of r people from a set of n of them, this argument shows that

$$C(n, r) = C(n-1, r-1) + C(n-1, r).$$

There is also an explicit formula: $C(n, r) = \frac{n!}{(n-r)!r!}$. This comes about by first observing that the number of permutations of n objects taken r at a time is $n \times (n-1) \times (n-2) \times \cdots \times (n-r+1) = \frac{n!}{(n-r)!}$, as there are n choices for the first object, followed by

$n - 1$ for the second, and so on until we have formed a row of r objects, whence the final multiplier is $(n - (r - 1)) = n - r + 1$. Any set of r distinct objects gives rise to $r!$ of these permutations, so that $C(n, r)$ is given by dividing this fraction by $r!$.

A subset of a set of size n can be coded as a binary string of length n. We consider the set in question in a specific order, $\{a_1, a_2, \cdots, a_n\}$ and then a binary string of length n specifies a subset by saying that each instance of 1 in the string indicates the presence of the corresponding a_i in the subset in question. For example, if $n = 4$ the strings 0111 and 0000, stand respectively for $\{a_2, a_3, a_4\}$, and for the empty set. Since there are two choices for each entry in the binary string there are 2^n such strings, and therefore 2^n subsets in all of a set of size n. Since the row labelled n of the Arithmetic Triangle counts all the possible subsets of a set of size n, (with the first row corresponding to $n = 0$), the sum of any row is 2^n, $n = 0, 1, 2, \cdots$.

Note 11 Page 68 *Diagonal sums give the Fibonacci numbers f_n.*

The claim that is being made here is that

$$f_{n+1} = C(n, 0) + C(n - 1, 1) + C(n - 2, 2) + \cdots$$
$$\text{for } n = 0, 1, 2, \cdots$$

We prove this by induction on n, the result being evidently true for $n = 0, 1$. For $n \geq 2$ we use the definition of f_{n+1} together with the recursive identity for binomial coefficients of the previous note and the fact that $C(n, 0) = 1$ for any value of n to argue that

$$f_{n+1} = f_n + f_{n-1} = (C(n - 1, 0) + C(n - 2, 1) + \cdots)$$
$$+ (C(n - 2, 0) + C(n - 3, 1) + \cdots)$$
$$= C(n - 1, 0) + (C(n - 2, 1) + C(n - 2, 0))$$

$$+ (C(n - 3, 1) + C(n - 3, 0)) + \dots =$$

$$C(n, 0) + C(n - 1, 1) + C(n - 2, 2) + \cdots \text{ as claimed.}$$

Note 12 Page 72 *Golden Rectangle* (See Note 14 below)

Note 13 Page 74 *Recursion for Stirling Numbers*

We argue similarly to that for the recursion for the binomial coefficients. In order to form a partition of a set of size n into r non-empty blocks we may proceed in two ways. We may take the first $n - 1$ elements of the set and partition it into $r - 1$ non-empty blocks in $S(n - 1, r - 1)$ ways, and the final member of the set will then form the rth block. Alternatively we may partition the first $n - 1$ elements of the set into r non-empty blocks, which can be done in $S(n - 1, r)$ ways, and then decide in which of the r blocks to place the final member of the set, giving us r choices. Hence we infer that

$$S(n, r) = S(n - 1, r - 1) + r S(n - 1, r) \text{ for } n = 1, 2, \cdots$$

Using this recursion formula we may calculate each line of the Stirling Triangle from the one above it.

We can compute $S(n, 2)$ and $S(n, n - 1)$ by inspection. An arbitrary partition of the n-set into a first set and a second set is described by a binary string of length n (see Note 10), where the presence of a 1 indicates membership of the first set. There are therefore 2^n such ordered pairs of sets. Since there is no ordering of the blocks within a partition we divide this number by 2 to find the number of partitions of the n-set into 2 sets, giving the number 2^{n-1}. However we need to subtract 1 from this in order to exclude the case where one of the sets is empty, hence $S(n, 2) = 2^{n-1} - 1$ for $n = 2, 3, \cdots$.

At the other extreme, a partition of the n-set into $n - 1$ blocks is determined by a choice of the unique block of size 2. The number of ways of making this selection is $C(n, 2) = \frac{1}{2}n(n - 1)$, the $(n - 1)$st triangular number.

Note 14 Page 74 *Binet's formula for the Fibonacci Numbers.*

An explicit formula for f_n can be found using a standard technique for solving so called *linear difference equations* or *linear recursions* of this type modelled on the technique used to solve linear differential equations with constant coefficients. We search for a solution to $f_n = f_{n-1} + f_{n-2}$ of the form $f_n = c^n$ for some unknown constant c. Substituting into the recursion gives that $c^n = c^{n-1} + c^{n-2} \Rightarrow c^2 = c + 1$. The solutions to this are the numbers $\alpha = \frac{1+\sqrt{5}}{2}$ and $\beta = \frac{1-\sqrt{5}}{2}$. We then seek to satisfy the initial conditions that $f_0 = 0$, $f_1 = 1$ (it is easier to start with f_0) by taking $f_n = a\alpha^n + b\beta^n$. Putting $n = 0, 1$ in turn gives the two equations $a + b = 0$, $a\alpha + b\beta = 1$ whence $a = \frac{1}{\sqrt{5}}$, $b = -\frac{1}{\sqrt{5}}$. Hence we find

$$f_n = \frac{1}{\sqrt{5}}\left(\frac{1+\sqrt{5}}{2}\right)^n - \frac{1}{\sqrt{5}}\left(\frac{1-\sqrt{5}}{2}\right)^n, \text{ for } n = 0, 1, 2, \cdots$$

This preposterous formula is of little use in calculating Fibonacci Numbers but it does give theoretical results. For instance, by looking at the ratio $\frac{f_{n+1}}{f_n}$ and taking the limit it is easy to see the observation of Kepler that the ratio approaches $\frac{1+\sqrt{5}}{2}$, which is called the Golden Ratio.

This number is also the length of the long side of the Golden Rectangle (Figure 4.4) for if we take the shorter side to be of unit length and we label the longer side by τ, the definition of the rectangle gives the equality of ratios: $\frac{\tau}{1} = \frac{1}{\tau-1}$ which is to say $\tau^2 = \tau + 1$, and we see that $\tau = \alpha$ as above.

As for the Lichtenberg ratio, take a rectangle of dimensions 1 and $\sqrt{2}$ and fold it down its long side. The two smaller rectangles appearing now have ratio of long to short sides of $\frac{1}{(\sqrt{2}/2)} = \frac{2}{\sqrt{2}} = \sqrt{2}$, and so the overall shape is maintained, and this process can now be repeated with the shape of the sheets remaining invariant.

Note 15 Page 75 *Ordered partitions and Fibonacci numbers.*

Let a_n be the number of ordered partitions of the non-negative integer n into integers exceeding 1. We see that $a_0 = 0$, $a_1 = 0$, $a_2 = 1$, $a_3 = 1$, $a_4 = 2, \cdots$. The claim is that for $n \geq 1$, $a_n = f_{n-1}$, the $(n-1)$st Fibonacci number, which is evidently true as far as we have listed the sequence of a_n. By considering the final number in an ordered partition of n of the required kind (which is either $2, 3, \cdots,$ or n) we infer that for $n \geq 2$ the a_n satisfy the recurrence:

$$a_n = a_{n-2} + a_{n-3} + \cdots + a_0$$

Next we note that the sum of all the terms of this recurrence on the right hand side apart from the first is, because of the validity of the recurrence itself, equal to a_{n-1} for all $n - 1 \geq 2$, that is for all $n \geq 3$. It follows that for all $n \geq 3$ the numbers a_n satisfy the Fibonacci recurrence $a_n = a_{n-2} + a_{n-1}$, from which the claim follows.

Note 16 Page 80 *Long lists of composite numbers*

For example, a list of n consecutive numbers that are all composite is given by $(n + 1)! + 2$, $(n + 1)! + 3$, $(n + 1)! + 4, \cdots, (n + 1)! + n$, $(n + 1)! + n + 1$; the first is divisible by 2, the second by 3, and so on, giving a list of n numbers, none of which are prime.

Note 17 Page 80 *An infinity of primes of the form $6n - 1$.*

Consider the list of primes $2, 3, \cdots, p$ and put $q = (2 \cdot 3 \cdots p) - 1$, which has the form $6n - 1$. Now all prime divisors of q exceed p. These prime factors cannot *all* be of the form $6n + 1$, for then q would be too. Therefore there is at least one prime r of the form $6n - 1$ with $p < r \leq q$.

Note 18 Page 83 *Formulas for primes*

Let $f(x) = a_0 + a_1x + \cdots + a_kx^k$ be a non-constant polynomial and suppose that $f(a) = y \geq 2$. Then $f(x)$ cannot always be prime because $f(a + ry)$ has y as a factor:

$$f(a + ry) = a_0 + a_1(a + ry) + \cdots + a_k(a + ry)^k = f(a)$$

$$+ \text{ terms in powers of } y.$$

Since y divides all terms on the right hand side, it follows that y is a factor of $f(a + ry)$ for all $r = 0, 1, 2, \cdots$ and since these numbers $f(a + ry)$ cannot all equal y, some of them must be composite.

The recursion tested in the text is given by $a_1 = 1$, and $a_n = 2a_{n-1} + 1$ for all $n = 2, 3, \ldots$ From this it is readily proved by induction that $a_n = 2^n - 1$ for all n. In general formulas of the form $a^n \pm 1$ all run into trouble as formulas for primes because they are subject to certain factorizations. In the case of the minus sign we have the expression that is the basis of summing a geometric progression:

$$a^n - 1 = (a - 1)(a^{n-1} + a^{n-2} + \ldots + 1).$$

This shows that $a^n - 1$ cannot be prime unless $a = 2$. Even in this case however, if $n = ab$ is a composite number we have the factorization:

$$2^n - 1 = (2^a - 1)(2^{a(b-1)} + 2^{a(b-2)} + 2^{a(b-3)} + \ldots + 2^a + 1)$$

and so $2^n - 1$ is also composite. For example if $n = 15 = 3 \times 5$ we get upon putting $a = 3$ and $b = 5$ that

$$32,767 = 2^{15} - 1 = (2^3 - 1)(2^{12} + 2^9 + 2^6 + 2^3 + 1) = 7 \times 4,681.$$

The problem of the primality of the numbers of the form $2^n - 1$ has therefore been reduced to the problem of the *Mersenne numbers* $2^p - 1$, where p is a prime. Although this sequence is rich in primes, some Mersenne numbers such as $2^{11} - 1 = 2,047 = 23 \times 89$ are not prime. All the same it can be proved that any proper factor of a Mersenne number has the form $2kp + 1$. In the case of $p = 11$ this tells us that only factors of the form $22k + 1$ are possible, and the two actual factors arise through the values $k = 1$ and $k = 4$ respectively. This property makes the Mersenne numbers prime candidates, in the strict sense of the word, and that is why they are the basis of the search for extremely large primes. As mentioned in the first chapter, at the time of writing, the 44th Mersenne prime is the World Champion.

The Mersenne primes are famously linked to the search for *perfect numbers*, which are the numbers that equal the sum of their own factors, as Euclid proved (2000 years before Mersenne's time) that given any Mersenne prime, $2^p - 1$, the number $2^{p-1}(2^p - 1)$ is an even perfect number. The values of $p = 2, 3, 5$ yield the first three perfect numbers, 6, 28, and 496 respectively. Euler showed the converse, that all even perfect numbers arise through Mersenne primes and this formula. This reduces the problem of finding even perfect numbers to that of finding Mersenne primes. However, we do not even know if there are infintely many Mersenne primes. Nor do we know if there are any odd perfect numbers. There have been strings of results that put extraordinary restrictions on any odd perfect numbers, without managing to legislate them out of existence. For instance Euler knew that any odd perfect number has the form $p^{4k+1}Q^2$, where p is a prime of the form $4n+1$. In 2005, Hare showed that an odd perfect must have at least 75 prime factors and Neilsen proved a year later that it must have at least nine distinct primes dividing it. We know there is no odd perfect less than 10^{300}, and there is more news along these lines to be found on the *Wolfram's World of Mathematics* web page.

As regards a number of the form $a^n + 1$, we see that this is even if a is odd, thus leaving only the case where a is an even number. If however n has an odd factor so that $n = mt$ say with m odd, then we have a peculiar telescoping factorization:

$$a^n + 1 = (a^t + 1)(a^{(m-1)t} - a^{(m-2)t} + a^{(m-3)t} - \cdots + 1)$$

For example, if $a = 2$ and $n = 11$ so that $m = 11$ and $t = 1$ we have $2^{11} + 1 = 2,049 = (2^1 + 1)(2^{10} - 2^9 + 2^8 - \cdots + 1) = 3 \times 683$. We do need the m to be odd in this factorization in order that the alternating pattern of plus and minus signs terminates with a $+1$, which in turn is necessary to ensure that the right-hand side does simplify correctly to the left-side. We conclude that $a^n + 1$ cannot be a formula for primes unless a is even and n itself is a power of 2. For $a = 2$ the numbers in question are those of the form $F_n = 2^{2^n} + 1$ and are known as the Fermat numbers.

These numbers are especially significant for it was proved by Gauss when still a teenager that a regular polygon is constructible with compasses and a staightedge if and only if the number of sides is a Fermat *prime* multiplied by some power of 2. (The powers of 2 arise because, give a regular polygon, we can always bisect all the sides, multiplying the possible side number by 2, then 4, then 8, and so on.) For $n = 1$ we have the regular pentagon, which features in *Euclid's Elements*, the construction itself being based on construction of the Golden Ratio. Gauss celebrated his proof by explicitly constructing the 17-gon, which is the $n = 2$ case. The feat of constructing F_3, the 257-gon was accomplished by Richelot and Schwendenwein in 1832 and J. Hermes spent ten years on the next one, the 65,537-gon and deposited his effort in a box at the University of Gottingen where it may still lie. Every Fermat number past $n = 4$ whose primality has been settled has turned out to be composite, so they give no more constuctible regular polygons, and the list goes at least as far as the enormous F_{16}. It seem safe to say then that the efforts of Hermes will never be surpassed!

That F_5 is not prime was settled by Euler who somehow spotted that $F_5 = 2^{32} + 1 = 4,294,967,297 = 641 \times 6,700,417$, thereby showing that the Fermat numbers do not represent a formula for primes. It is easy to check that 641 is prime (it has no prime factor up to 23, which exceeds its square root) and the hint that it could bear a special relationship to F_5 comes by way of the facts that $641 = 2^4 + 5^4 = 5 \times 2^7 + 1$. Arithmetic modulo a *prime* represents a finite field in which we can add, subtract, multiply and importantly divide as in ordinary arithmetic, across the modular equivalence sign. This then allows the following remarkable sequence of manipulations modulo 641:

$$2^4 + 5^4 \equiv 0 \Rightarrow \frac{2^4}{5^4} + 1 \equiv 0 \Rightarrow \frac{2^4}{5^4} \equiv -1 \,(\mathrm{mod}\,641).$$

While the second relationship can be expressed as:

$$5 \times 2^7 + 1 \equiv 0 \Rightarrow 5 \times 2^7 \equiv -1 \Rightarrow 2^7 \equiv -\frac{1}{5} \,(\mathrm{mod}\,641).$$

Multiplying both sides of this by 2 we obtain:

$$2^8 \equiv -\frac{2}{5} \,(\mathrm{mod}\,641);$$

and raising both sides to the power 4 then reveals that

$$2^{32} \equiv \left(-\frac{2}{5}\right)^4 = \frac{2^4}{5^4} \pmod{641};$$

but then our first equation allows us to say that $2^{32} \equiv -1$ $\pmod{641}$, in other words:

$$F_5 = 2^{32} + 1 \equiv 0 \pmod{641};$$

or in other words, 641 is a factor of the fifth Fermat number, which is therefore not prime.

Note 19 Page 83 *Fermat's little lemma*

Fermat's Little Lemma is that both a and a^p leave the same remainder when divided by the prime p. Here is a proof quite different from the standard one. Imagine we have a types of colored beads, so there are a^p different rows of beads of length p that we can make. Consider the $b = a^p - a$ rows that do *not* consist of only one color bead. Call two rows equivalent if they are identical when formed into a necklace, that is a circle of beads. Now each row is equivalent to no more than p other rows—in fact exactly p other rows. One necklace can only arise from at most p different rows, corresponding to the p places we can break the necklace to form a row. On the other hand, two of the rows formed by breaking the necklace can only be the same if the necklace consists of m copies of the same string of beads of length n say, so that $mn = p$. For example writing B for blue and Y for yellow, the following six strings would all form the same necklace:

$$BYBYYB,\ YBYYBB,\ BYYBBY,\ YYBBYB,$$
$$YBBYBY,\ BBYBYY.$$

We form each string from the previous one, by taking the first bead on the left and placing it at the right hand end—this also applies to the first string, as that is the result of taking the final string and doing the same. These six strings are therefore equivalent. However, suppose we begin with the string $BYYBYY$. Taking different starting places in the string, the variants that result number only three:

$$BYYBYY, YYBYYB, YBYYBY;$$

as the next variant takes us immediately back to $BYYBYY$. This possibility only arises if the string consists of a short block that is repeated a number of times. However, *since p is prime,* this case does not arise, as we must have *m = 1 or m = p, and since the necklace is not monochromatic, m ≠p*. It follows that *m = 1*, and that p is a factor of $b = a^p - a$, because the b rows can be grouped into disjoint sets of size p. Therefore a^p and a leave the same remainder when divided by the prime p.

The standard proof exploits the nature of the product $(p - 1)!$ modulo p, while another is by induction on a and makes use of the factors of binomial coefficients. Both are also quite short.

Chapter 5

Note 20 Page 96 & 97 *Geometric and harmonic means; Heron's formula for roots*

$$\frac{1}{H} = \frac{1}{2}\left(\frac{1}{a} + \frac{1}{b}\right) \Rightarrow \frac{2}{H} = \frac{b + a}{ab} \Rightarrow H = \frac{2ab}{a + b}.$$

Next we show that $G \leq A$. We begin with the observation that

$$(\sqrt{a} - \sqrt{b})^2 \geq 0 \Rightarrow a + b - 2\sqrt{ab} \geq 0 \Rightarrow A \geq G$$

with equality only when $a = b = A = G$.

Similarly we may show that $H \leq G$, by starting with $(a - b)^2 \geq 0 \Rightarrow$

$$a^2 + b^2 \geq 2ab \Rightarrow (a + b)^2 \geq 4ab \Rightarrow ab(a + b)^2 \geq 4a^2b^2 \Rightarrow$$

$$ab \geq \frac{4a^2b^2}{(a + b)^2}, \text{ which, upon taking square roots gives } G \geq H.$$

Again we have strict inequality, except when all quantities involved are one and the same.

The Heron iteration for finding $\sqrt{2}$ is to make a first guess, $a_0 = a$ and then compute the sequence of values, $a_n = \frac{1}{2}(a_{n-1} + \frac{2}{a_{n-1}})$ for successive values of n. This gives an excellent approximation to $\sqrt{2}$ in a few iterations from any positive starting value. It is itself just a special case of the Newton-Raphson method applied to the function $y = x^2$, whereby a root of the given function is found by an iteration that itself arises from approximating the root to the x−intercept of the tangent to the curve from a given starting point. That the Heron method does converge can be seen by putting $a = a_{n-1}$ and $b = \frac{2}{a_{n-1}}$ in the inequality $H \leq G \leq A$ for in this case we obtain:

$$\frac{4}{a_{n-1} + \frac{2}{a_{n-1}}} \leq \sqrt{2} \leq \frac{a_{n-1} + \frac{2}{a_{n-1}}}{2} \Rightarrow \frac{2}{a_n} \leq \sqrt{2} \leq a_n$$

It follows that $a_n \geq \sqrt{2}$ for all $n \geq 1$ (irrespective of the value of the seed number a_0). Since a_{n+1} is the mean of the two values at either end of this inequality it follows that $a_n \geq a_{n+1} \geq \sqrt{2}$. We infer that at each stage the distance of the new approximation a_{n+1} to $\sqrt{2}$ is less than half the distance of the old approximation, a_n, and so the limiting value of the sequence of Heron iterations is indeed $\sqrt{2}$.

Chapter 6

Note 21 Page 105 $(-1) \times (-1) = 1$

We give a formal derivation of this fact as a consequence of the standard Laws of Algebra. We shall assume in particular that the integers satisfy:

Commutativity of Addition and Multiplication: $a + b = b + a$, $ab = ba$;

Associativity of Addition and Multiplication: $a + (b + c) = (a + b) + c, a(bc) = (ab)c$;

Distributivity of Multiplication over Addition: $a(b + c) = ab + ac$.

We assume that 0 is the *additive identity*, meaning that $a + 0 = a$ is always true, and that for any number a, there is a unique *opposite* or *additive inverse* as it is also called, denoted by $-a$, that has the property that $a + (-a) = 0$. Moreover we assume that 1 is the *multiplicative identity* of the number system so that $a \times 1 = a$ is always true.

Having established reasonable ground rules, we shall use them freely. We next need to prove that for any number a, $a \times 0 = 0$. To this end, let us write b for $a \times 0$.

$$b = a \times 0 = a \times (0 + 0) = a \times 0 + a \times 0 = b + b.$$

Now

$$b = b + b \Rightarrow b + (-b) = (b + b) + (-b)$$
$$\Rightarrow 0 = b + (b + (-b)) = b + 0 = b.$$

That is to say, $b = a \times 0 = 0$, as claimed. We now apply this additional fact to infer that $(-1) \times 0 = 0$ so that

$$0 = (-1) \times 0 = (-1) \times ((-1) + 1)$$
$$= (-1) \times (-1) + (-1) \times 1 \Rightarrow (-1) \times (-1) + (-1) = 0.$$

We now add 1 to both sides of this equation, and use associativity of addition to re-bracket in order to gain the required conclusion:

$$(-1) \times (-1) = 1.$$

Note 22 Page 107 & footnote *Egyptian Fractions and the Akhmim papyrus*

Take any proper non-unit fraction $\frac{m}{n}$, in reduced form, and write $n = km + r$ for some $1 \le r \le m - 1$ and $k \ge 1$. (Note that both r and k are at least 1 as the fraction is reduced and $0 < \frac{m}{n} < 1$, with $m \ge 2$.) The largest reciprocal less than $\frac{m}{n}$ is then $\frac{1}{k+1}$ as

$$km < n = km + r < km + m = m(k + 1).$$

Taking reciprocals throughout (which causes the inequalities to change direction) gives:

$$\frac{1}{km} > \frac{1}{n} > \frac{1}{m(k + 1)} \Rightarrow \frac{1}{k + 1} < \frac{m}{n} < \frac{1}{k}.$$

That is to say that $\frac{1}{k+1}$ is the largest unit fraction less than $\frac{m}{n}$, as the next largest, $\frac{1}{k}$, is too big. Now look at what happens when we subtract:

$$\frac{m}{n} - \frac{1}{k + 1} = \frac{m(k + 1) - n}{n(k + 1)} = \frac{m(k + 1) - (mk + r)}{n(k + 1)}$$

$$= \frac{mk + m - mk - r}{n(k + 1)} = \frac{m - r}{n(k + 1)}.$$

The key observation is that the numerator has reduced from m, to $m - r$. Since r is positive, this is a decrease, and since $r < m$, the final numerator is also positive. Hence the sequence of numerators of the remaining part of the fraction is a decreasing sequence of positive integers, so that after $m - 1$ steps or fewer, the numerator will become 1.

It remains only to note that the next reciprocal subtracted will always be smaller than the previous one (as this will ensure the additional requirement that the unit fractions in the decomposition are all different, is met). By the way it is selected, we see that the next reciprocal cannot be larger than its predecessor, and nor can it be equal as

$$\frac{1}{k + 1} + \frac{1}{k + 1} = \frac{2}{k + 1} > \frac{m}{n} \text{ as } \frac{m}{n} < \frac{1}{k} \le \frac{2}{k + 1},$$

as this last inequality is equivalent to saying that $k + 1 \le 2k \Leftrightarrow 1 \le k$, which is true, as $\frac{m}{n}$ is proper, so that $m < n$.

For example, applying this procedure to $\frac{9}{20}$, we get $20 = 2 \times 9 + 2$, so $m = 9, n = 20, k = 2 = r$. This gives $\frac{1}{k+1} = \frac{1}{3}$. We subtract accordingly to get $\frac{9}{20} - \frac{1}{3} = \frac{7}{60}$ and, in accord with the general algebraic description, the numerator has decreased from 9 to 7. We repeat the algorithm with $m = 7, n = 60$: $60 = 8 \times 7 + 4$, so that we subtract $\frac{1}{9}$ to give $\frac{7}{60} - \frac{1}{9} = \frac{1}{180}$. We arrive at the Egyptian decomposition $\frac{9}{20} = \frac{1}{3} + \frac{1}{9} + \frac{1}{180}$.

This method always works, but the superior decomposition $\frac{9}{20} = \frac{1}{4} + \frac{1}{5}$ can be found using a technique of a Greek papyrus discovered at the city of Akhmim on the Nile and dated to 500–800 AD. In modern notation the trick can be expressed as the readily verified algebraic identity:

$$\frac{m}{pq} = \frac{m}{p(p+q)} + \frac{m}{q(p+q)}.$$

Applying this with $m = 9, p = 4, q = 5$ immediately gives us $\frac{9}{20} = \frac{9}{4\times9} + \frac{9}{5\times9} = \frac{1}{4} + \frac{1}{5}$.

As a second example, let us decompose $\frac{2}{99} = \frac{2}{9} \times \frac{1}{11}$. Applying the Akhmim Technique to $\frac{2}{9}$, we take $m = 2, p = 1, q = 9$, and get $\frac{2}{9} = \frac{2}{1\times10} + \frac{2}{9\times10} = \frac{1}{5} + \frac{1}{45}$, yielding the decomposition $\frac{2}{99} = \frac{1}{55} + \frac{1}{495}$.

In this case the greedy technique of always subtracting the largest reciprocal yields the decomposition $\frac{2}{99} = \frac{1}{50} + \frac{1}{4950}$, and there are other decompositions into the sum of a pair of unit fractions.

Note 23 Page 111 footnote *The Pigeonhole Principle*

Although a simple minded idea, versions of the Pigeonhole Principle are used time and again to prove results about inevitability in mathematical structures, both finite and infinite. Here are two examples. Any set of $n + 1$ numbers from the first $2n$ positive numbers must contain a number that is a factor of one of the others. We can see that this claim fails if we replace $n + 1$ by n in the statement by virtue of the set $n + 1, n + 2, \cdots, 2n$, which gives the required contradiction.

To verify the claim we begin by observing that any number m can be written in the form $m = 2^k t$, where $k \geq 0$ and t is odd.. The index k will be zero exactly if m is already odd and the number t will be 1 if m happens to be a power of 2 and not otherwise. Given that m lies in the range from 1 to $2n$, so does its odd factor t. However, there are only n distinct odd numbers in this range so that if follows, by the Pigeonhole Principle, that two different numbers from our set of size $n + 1$ share the same odd factor t. Call these numbers m_1 and m_2, so that we have $m_1 = 2^{k_1} t$ and $m_2 = 2^{k_2} t$ say. The smaller of these two numbers, m_1 and m_2, is then a factor of the other, as required.

For our second trick, we show that given any eight numbers, the sum or difference of some pair of them must be a multiple of 13. Again, this is a sharp result as it is not true if we have just seven numbers as is seen through the example, 0, 1, 2, 3, 4, 5, 6.

We may assume that no two of the eight integers are congruent modulo 13, for otherwise their difference immediately gives us a required multiple of 13. In particular then, there are seven numbers, $b_1, b_2, \cdots, b_7$, which are not divisible by 13 and let a denote the eighth number to hand. Consider the numbers $a \pm b_i$, $(1 \leq i \leq 7)$. There are 14, (not necessarily distinct) numbers here and so, by the Pigeonhole Principle, for two subscripts, i, j, we have $a \pm b_i \equiv a \pm b_j$ (modulo 13). (Note that the $\pm$ signs are independent, that is to say they may or may not differ on either side of the $\equiv$ sign.) If $i = j$ we would obtain that $2b_i$, and hence b_i, is divisible by 13, contrary to the choice of the integers b_i. Hence $i \neq j$, and so either $b_i + b_j \equiv 0$ (modulo 13) or $b_i - b_j \equiv 0$ (modulo 13), in accord with our claim.

If you wish to try a problem of this type consider this claim: given any set of $n + 1$ numbers from $1, 2, \cdots, 2n$ at least two of them have no common factor.

Note 24 Page 112 *Converting recurring decimals to fractions*

The idea is best seen through examples, but can be described as follows. We are given a recurring decimal expansion of a number a. If the length of the recurring block is n, we multiply a by 10^n. If we then compute $10^n a - a$, the recurring parts will cancel. If we solve this equation for a, we recover a as a vulgar fraction, which may then be cancelled down to the extent possible. For example, let $a = 0.6\,\overline{81}\,\ldots$. The length of the recurring block is 2, so we calculate

$$100a - a = 99a = 68.1\overline{81} - 0.6\overline{81} = 67.5 \Rightarrow 990a$$
$$= 675 \Rightarrow a = \frac{675}{990} = \frac{15}{22}.$$

Note 25 Page 113 *Non-recurring decimals are irrational*

This observation is often the basis for clever exercises such as show that for any $k \geq 2$, $\sum_{n=1}^{\infty} k^{-n(n+1)}$ is irrational. This is immediate, as if we represent this number as a base k expansion, we have a non-recurring (binary) sequence $0 \cdot 0100010 \cdots$.

Note 26 Page 114 *Side length of a square is $\sqrt{2}$ without Pythagoras*

First imagine a square of side length 2 and divide it into four unit squares by drawing a cross in the middle. Consider the square formed by the four diagonals of the small squares that run between the midpoints of the sides of the original square. The area of the big square is 4. Since each diagonal splits its small square into two identical triangles, we see that the small square covers exactly half the big square, and so its area is 2. The side s of the small square forms the hypotenuse of an isosceles right triangle whose sides are of unit length. We have just in effect noted that $s^2 = 2$ so that the side of the diagonal of each small square is indeed $\sqrt{2}$. This Ancient Indian derivation of this key fact does not appeal to Pythagoras's Theorem.

Note 27 Page 115 *nth roots are irrational ($n \geq 2$)*

Suppose that $k^{\frac{1}{n}} = \frac{a}{b}$, where all symbols represent positive integers. We shall show that k is then an *nth* power, $k = t^n$ say, whence

$\frac{a}{b} = t$ is an integer. We do make implicit use of the Fundamental Theorem of Arithmetic (FLT), which says that the prime factorization of a number is unique.

Our equation gives $a^n = kb^n$. Let p be any prime factor of k, so that $k = p^m l$ say, where p is not a factor of l. Since a^n, b^n are both nth powers, then $a^n = p^{rn}c$, $b^n = p^{sn}d$ say for some non-negative integers r and s and integers c and d, which are not multiples of p. Equating powers of p on both sides of the first equation (which is justified by the FLT) now gives us $rn = sn + m \Rightarrow m = n(r - s)$. It follows that the highest power m of p that divides k is a multiple of n. Since this applies to every prime factor p of k, it follows that k is itself an nth power, as required.

Therefore the nth root of a positive integer is either another integer or is irrational.

Chapter 7

Note 28 Page 124 *The set of algebraic numbers is countable*

We need the fact that a polynomial $p(x) = a_0 + a_1x + a_2x^2 + \cdots + a_nx^n$ has at most n roots, that is to say that there at most n solutions to the equation $p(x) = 0$. This is proved by induction on n, the result being clear for $n = 1$, so suppose that $n \geq 2$. If $p(x)$ has a root r then, by the Factor Theorem, $p(x) = (x - r)q(x)$, where $q(x)$ is a polynomial of degree $n - 1$ which, by induction, has at most $n - 1$ roots. If $p(a) = 0$, either $a - r = 0$, whence $a = r$, or $q(a) = 0$, whence a is one of the roots of $q(x)$. Therefore $p(x)$ has at most $1 + (n - 1) = n$ roots, and the induction continues.

Now suppose that a is an algebraic number, so that $p(a) = 0$ for some polynomial of degree n say, with rational coefficients, which we take to have the form indicated above. By multiplying the equation $p(a) = 0$ by the product of the denominators of all the coefficients a_i, we find that a is a root of a polynomial of the same degree with integer coefficients, so that we conclude that the algebraic numbers are exactly those that are roots of polynomials with integer coefficients. We shall denote by P this set of polynomials.

Now let P_n be the set of all those members of P of degree no more than n, whose coefficients a_i satisfy $-n \le a_i \le n$. Note that each P_n is a finite set of polynomials. (Indeed there are $2n + 1$ choices for the value of each a_i, and $n + 1$ coefficients to choose, so that $|P_n| = (2n + 1)(n + 1)$.) Let A_n be the set of numbers that are roots of some polynomial in P_n. Since each member of P_n has at most n roots, the size of A_n is also finite. (Indeed $|A_n| \le n(n + 1)(2n + 1)$.) Since every algebraic number is in some A_n, we may now form a list of all algebraic numbers by taking in turn all the members of the (finite) set A_1, then all the members of A_2 (not repeating any already in A_1 if we wish), then all the members of A_3, and so on.

Therefore the set of all algebraic numbers is a countable set.

It follows that the set T of all transcendental (i.e. non-algebraic) numbers is uncountable, for if T were countable, then we could list all the members of the union $S \cup T$ of these two sets, by taking the lists for S and T respectively and interleaving them, taking one member from S, then one from T, then one from S, then another from T, and so on. However since $S \cup T$ is the set of all real numbers, which is uncountable by Cantor's Argument, if follows that the set of all transcendental numbers is indeed an uncountable collection.

Note 29 Page 125 *Set pairing: Schröder-Bernstein Theorem*

It should be appreciated that, as for finite sets, we may say the number of one set A is no greater than another B if there is a one-to-one mapping from A into B. For this to be a sensible definition applying to infinite sets, we require that if A is no greater than B in this sense, and if B is also no greater than A, then the sets have the same cardinality. This is however not obvious, and needs proof. The theorem we require is that if there is one-to-one mapping from A into B, and another one-to-one mapping from B into A, (whose ranges might only be subsets of B and of A respectively) then there is a full one-to-one correspondence between the elements of A and those of B. This fact is known as the Schröder-Bernstein Theorem, and is one of the basic results proved in a course on the general theory of sets.

Note 30 Page 125 *Cardinality of the set of subsets*

Take for example the subset of the positive integers consisting of all odd numbers. We may 'code' this set as an infinite string of 0's and 1's: in this case the relevant string is 101010 ... where a 1 indicates that a number is present in the subset under discussion, and the 0 denotes the opposite (cf Note 10). This string therefore is indicative of the subset consisting of 1, 3, 5, etc. as the 1's appear in the first, third, and fifth places etc. In a similar way the string corresponding to the primes would begin 01101010001 ..., indicating the presence of the numbers 2, 3, 5, 7, 11 etc. We don't know exactly how this string goes because we don't know what all the prime numbers are, but there is a string that corresponds to the prime numbers just as there is for every other subset of the counting numbers. Even finite subsets are included: for example the subset consisting of just the number 2 has for its string $01000\cdots$.

The trick now is to remind ourselves that any string of 0's and 1's can be taken as a number between 0 and 1 written in binary (base 2) if we slip a decimal point in front of the lot. (For example, the string for the set {2} just given corresponds to the number $\frac{1}{4}$, as the solitary 1 in the expansion would stand for $\frac{1}{2^2}$.) If you are not too happy working in binary arithmetic, and few of us are, have no fear, for the only reason that it is mentioned at this juncture is to allow us the conclusion that there *is* a one-to-one correspondence between the collection of all subsets of the counting numbers and the set of all numbers between 0 and 1. The binary representation is merely a device, a trick if you like, that allows us to see this is true. The correspondence we have set up is an arbitrary one that does not have any real meaning or interest in itself, but it is a pairing of the sets nonetheless. Since the set of all real numbers between 0 and 1 is uncountable, so is the set of subsets of the counting numbers. This then is an infinite example of the general phenomena by which the collection of all subsets of a set is always larger than the set itself, in the sense that the two collections cannot be paired off, one against the other. The set of subsets is just too big—it has a higher cardinality.

Note 31 Page 127 *e is irrational*

We use the representation $e = 1 + \frac{1}{1!} + \frac{1}{2!} + \frac{1}{3!} + \cdots$ and exploit the fact that this series converges very quickly to show that a rational limit is impossible. Suppose to the contrary that, $e = \frac{p}{q}$ for some positive integers p and q. Partition this series into two parts, the sum of the terms as far as $\frac{1}{q!}$ forms the first part, and the remainder is the second part:

$$\frac{p}{q} = \left(1 + \frac{1}{1!} + \cdots + \frac{1}{q!}\right) + \left(\frac{1}{(q+1)!} + \frac{1}{(q+2)!} + \cdots\right).$$

Now multiply through by $q!$ to give:

$$p(q-1)! = \left(q! + \frac{q!}{1} + \frac{q!}{2} + \cdots + q + 1\right)$$
$$+ \left(\frac{1}{q+1} + \frac{1}{(q+1)(q+2)} + \cdots\right)$$

The left hand side of this equation is an integer, as is every term in the first bracket on the right. Hence the sum of the remaining term, since it equals the difference of two integers, is itself an integer. However we show this is impossible, by verifying that it is positive (obvious) and less than 1. The sum in question is clearly less than that which results through replacing each factor in each denominator by $q + 1$; and this is

$$\frac{1}{q+1} + \frac{1}{(q+1)^2} + \frac{1}{(q+1)^3} + \cdots = \frac{1}{q+1}\left(1 + \frac{1}{q+1}\right.$$
$$\left. + \frac{1}{(q+1)^2} + \cdots\right)$$

The series in brackets is an infinite geometric progression, and as such its sum is

$$\frac{1}{1 - \frac{1}{q+1}} = \frac{q+1}{q}$$

Hence the entire expression in question is equal to $\frac{1}{q} \leq 1$, and so the second bracketed term above is less than 1. We have reached the desired contradiction, so we deduce that e is not a rational number. Indeed it can be shown after considerably more work that e is transcendental.

Note 32 Page 127 *Matching card decks*

The comparison of the card decks (with n cards let us say) is in effect the choice of a permutation on n symbols. That no coincidence arises corresponds to the the permutation being a *derangement,* that is to say a permutation that has no fixed point. The total number of permutations is $n!$. The total number of derangements on the other hand is given by the following alternating sum:

$$\sum_{k=0}^{n} (-1)^k C(n, k)(n - k)!$$

as the first term counts all permutations, the second subtracts all those that fix at least one point, the second adds on all those that fix at least two points, and so on—overall this only counts permutations with no fixed points: for example, if a permutation has three fixed points, it is counted positively for $k = 0$ and $k = 2$ (a total of $1 + 3 = 4$ times, the 3 arises as there are 3 ways of choosing two fixed points from the nominated three) and negatively when $k = 1$ and $k = 3$ (a total of $3 + 1 = 4$ times), and so contributes 0 to the overall count. (What is working for us here is the symmetry of the rows of Pascal's Triangle); a derangement counts exactly once in the first term, and does not feature in the count of further terms. Dividing this number by $n!$, we see that the probability of a derangement for an n-card deck is:

$$\sum_{k=0}^{n} \frac{(-1)^k}{k!}$$

which is the first $n + 1$ terms of the series for e^{-1}. Since this series coverges very quicky, certainly for $n = 52$ its value is indistinguishable from $\frac{1}{e}$.

Note 33 Page 130 *Rationals versus irrationals.*

Suppose that c is rational, let t be any number, and put $d = t + c$. Then $t = d - c$, and if d were rational so would be $t = d - c$, as its value could be found by doing the corresponding subtraction sum. It follows that if t is irrational, then so must be $d = t + c$. For example, $\pi - \frac{22}{7}$ is irrational. In a similar way, ct is irrational if c is not zero. For example, $\frac{1}{3}\sqrt{10}$, is irrational, as is $\frac{e-1}{2}$.

It is perfectly possible however for the sum or product of two positive irrational numbers to be rational. Examples include $(2 - \sqrt{2}) + \sqrt{2} = 2$ and $\sqrt{2} \cdot \sqrt{8} = \sqrt{16} = 4$.

What is more surprising is that there must exist two irrational numbers, a and b, such that a^b is rational. We might first try putting $a = b = \sqrt{2}$. If a^b is rational, we have made the point. If not (which might look the more plausible alternative) then put $a = \sqrt{2}^{\sqrt{2}}$ and take $b = \sqrt{2}$. Then both a and b are irrational, yet $a^b = (\sqrt{2}^{\sqrt{2}})^{\sqrt{2}} = (\sqrt{2})^2 = 2$. In either case, we see there exist a pair of irrationals, a and b, as required. Unfortunately perhaps, the proof gives us no clue as to which of the pairs does the job. The argument just says if one pair does not work, then the other pair must!

Numbers such as $\log 3$ are irrational also, for if $\log 3 = \frac{a}{b}$, a rational fraction, then $10^{\frac{a}{b}} = 3$, whence $10^a = 3^b$, and something is plainly wrong as the left hand side is even while the right hand side is odd!

Chapter 8

Note 34 Page 150 *Invincible teams*

The given formula may be proved by induction, the first few cases already having been checked. Suppose that the formula gives the correct answer for an $(n-1)$-round competition, and consider an n-round contest. In order to get a Celtic-Rangers final, the teams certainly need to miss one another in the first round. The probability of this is $q = \frac{2^n - 2}{2^n - 1}$. Given that this has occurred, the competition has now been reduced to an $(n-1)$-round contest. The probability of that ending in a Celtic-Rangers final is p_{n-1}. Hence by induction

$$p_n = q p_{n-1} = \frac{2^n - 2}{2^n - 1} \cdot \frac{2^{n-2}}{2^{n-1} - 1} = \frac{2(2^{n-1} - 1)}{2^n - 1} \cdot \frac{2^{n-2}}{2^{n-1} - 1}$$
$$= \frac{2^{n-1}}{2^n - 1},$$

and so the induction continues and the formula is proved.

By dividing top and bottom by 2^{n-1}, it is now straightforward to check that

$$p_n = \frac{1}{2 - \frac{1}{2^{n-1}}};$$

and so $p_n \to \frac{1}{2}$ from above as $n \to \infty$. With a little more work we can show that the probability that Celtic and Rangers meet in round k $(1 \le k \le n)$ is given by $\frac{2^{k-1}}{2^n - 1}$. The sum of these probablilites over all values of k is of course 1 as the invincible teams, being invincible, must clash somewhere in the competition.

Note 35 Page 152 *Birthday Coincidence Problem*

We shall ignore the slight complication of leap years and will assume what is approximately true that birthdays occur randomly throughout the year. If we have n people, we ask first for the probability that they all have *different* birthdays. In effect we are making n choices from a set of 365 distinct objects with replacement, meaning that it is possible for a choice to be taken up more than once. For $n = 2$ the probability of different birthdays is simply $\frac{364}{365}$—the probability that the second choice differs from the first. For $n = 3$ the corresponding probability is $\frac{364}{365} \times \frac{363}{365}$, because of the proportion of times that the first two people have different birthdays, the proportion in which the third person differs from both is given by the second fraction. Continuing in this way, the probability for an arbitrary value of n, is a product of $n - 1$ fractions:

$$\frac{364}{365} \times \frac{363}{365} \times \frac{362}{365} \times \cdots \times \frac{(366 - n)}{365}.$$

As n increases, this probability gradually diminishes towards 0, and indeed reaches 0 when $n = 366$, when the final terms in the product is 0. This is clear and is an instance of the so-called *Pigeonhole Principle.* If we have more letters than pigeonholes, then at least one mail slot will get more than one letter. Therefore if we have 366 people, there must be a birthday coincidence, as there are more people than birthdays.

Our original question though was how large a value of n do we need to take before this probability of no coincidence falls below one half. This can only be found by direct computation. It turns out that the smallest value of n for which this is true is $n = 23$. If you have 23 or more people in a room then you have a better than 50-50 chance of at least one birthday coincidence. Since birthdays are not quite evenly spread throughout the year, the chances of a coincidence is even higher than this and so, in practice even smaller groups are likely to have people in them who share a birthday.

Note 36 Page 154 *Russian Roulette*

Let us suppose that the probability of success is p (so $p = \frac{1}{6}$ in our problem) and write q for $1 - p$. We suppose that A starts first, but then B and A alternate in having two attempts each after that until a success arises. Once again let a and b be the respective probabilities that A or B wins and, as before, we have $a + b = 1$. The probablity that B wins on his very first attempt is qp. Given that neither A nor B win on their first attempt, the probability that it is B that goes on to win is a, as their initial positions have, in effect, been reversed. Hence we infer that $b = qp + q^2 a$, and substituting this into the first equation yields

$$1 = a + (qp + q^2 a) = a(1 + q^2) + qp$$
$$\Rightarrow a = \frac{1 - qp}{1 + q^2} = \frac{1 - q + q^2}{1 + q^2} = 1 - \frac{q}{1 + q^2}$$

In particular, if $p = \frac{1}{6}$ so that $q = \frac{5}{6}$ we get, as claimed, $a = \frac{31}{61}$. Indeed it is always the case that $a \geq \frac{1}{2}$ as $\frac{q}{1+q^2} \leq \frac{1}{2}$ as the latter inequality can be readily worked to $(q - 1)^2 \geq 0$. When $q = 1$ we do get $a = b = \frac{1}{2}$ from the equation although, if $q = 1$ the gun will never go off and the equation $a + b = 1$ does not apply!

Note 37 Page 157 *Waiting time for the bus*

The exact average time can be calculated as follows. Suppose that the longer period is $1 + t$ hours, so that the shorter period has duration $1 - t$. The chance of the passenger hitting the longer interval, is the length of that interval, divided by the length of the total time period of 2 hours: $(1 + t)/2$. Similarly $(1 - t)/2$ represents the probability of her arriving during the shorter waiting period. Given that she has arrived during a particular interval, her expected waiting time is half of the the interval's length, which is also $(1 + t)/2$ is the former case, and $(1 - t)/2$ otherwise. Adding the two separate contributions to her expected waiting time from the two intervals gives the value:

$$\frac{1+t}{2} \cdot \frac{1+t}{2} + \frac{1-t}{2} \cdot \frac{1-t}{2} = \frac{1}{2}(1 + t^2).$$

We see that if t were equal to 0, (no disruption to arrival time), then the average time to wait is half an hour as we would expect. However, if there is any disruption at all to the arrival time of the bus, (corresponding to t not being zero), then the passenger's wait is expected to be longer than half an hour. At the other extreme, if $t = 1$, corresponding to one bus not turning up at all only for a pair of them to turn up together after another hour, then we have in effect a two-hour interval between buses, and the passenger who arrives during this time can expect to endure an average wait of one hour.

Note 38 Page 159 *Losing on the Lottery for 10,000 years*

This is more than likely. The question is, more or less, if Buffon has a one in a million chance of winning and enters 500,000 times in a row, what is the probability that he loses every time? In general, if the chance of winning is $\frac{1}{n}$ then the exact probability of losing $\frac{n}{2}$ times in succession is:

$$\left(1 - \frac{1}{n}\right)^{\frac{n}{2}} = \left(\left(1 - \frac{1}{n}\right)^{n}\right)^{\frac{1}{2}} \approx e^{-\frac{1}{2}} = 0.6065$$

It is more than 60% likely that Buffon will still be waiting in the year 12,008 for his lucky numbers to come up. What is more, all his bad luck has done him no good—someone who has given him ten millenia head start will now be just as likely as him to win in future and, most probably, one of them would, to Buffon's eternal frustration!

Chapter 9

Note 39 Page 179 *Solution of the Quadratic through Completing the Square*

Take the quadratic equation to have the form $ax^2 + bx + c = 0$. We may assume that $a \neq 0$, as otherwise we would have a linear equation only. Hence we may divide throughout by a, and take the constant term to the right hand side to get $x^2 + \frac{b}{a}x = -\frac{c}{a}$. The trick now is to *complete the square,* so that the left hand side can be written as a perfect square plus a constant: $(x + p)^2 + q$. Since $(x + p)^2 = x^2 + 2px + p^2$, we must have $2p = \frac{b}{a}$, that is $p = \frac{b}{2a}$ to make this possible. The first two terms, $x^2 + 2px$, then correspond exactly to the left hand side, while the extra term $p^2 = \frac{b^2}{4a^2}$, which we have introduced needs to be added to the right hand side also. Having done this, we have completed the hard work for the left hand side is now a perfect square:

$$x^2 + \frac{b}{a}x + \frac{b^2}{4a^2} = -\frac{c}{a} + \frac{b^2}{4a^2} \Rightarrow (x + \frac{b}{2a})^2 = \frac{b^2 - 4ac}{4a^2}.$$

It now remains only to take square roots (plus and minus of course) and tidy up in order to derive the famous quadratic formula:

$$x + \frac{b}{2a} = \frac{\pm\sqrt{b^2 - 4ac}}{2a} \Rightarrow x = \frac{-b \pm \sqrt{b^2 - 4ac}}{2a}.$$

Note 40 Page 181 *Solution of the Cubic: Viète's Substitution*

By dividing throughout by the leading coefficient, any cubic equation is equivalent to one of the form

$$x^3 + ax^2 + bx + c = 0.$$

Moreover, by substituting $x = y - \frac{a}{3}$, we get a cubic in y for which the coefficient of y^2 is zero. (In a similar way, the nth degree equation can be reduced to one in which the term in y^{n-1} is absent, by putting $x = y - \frac{a}{n}$.) If follows that any cubic can be solved as long as we can solve the special cubic equations that have the form $x^3 = px + q$, that is to say as long as we can determine the points where the standard cubic curve $y = x^3$ meets an arbitrary straight line.

Cardano and his contemporaries certainly realized this much, but it was making progress from here that proved difficult. The equation can be solved from this point however by means of the *Viète Substitution*, $x = w + \frac{p}{3w}$. This reduces the cubic to the equation $w^3 + \frac{p^3}{27w^3} - q = 0$. On multiplying through by $z = w^3$, we obtain the quadratic equation $z^2 - qz + \frac{1}{27}p^3 = 0$, which can then be solved to find z, and then w, and finally x.

Note 41 Page 182 *Rational Root Theorem*

If $\frac{p}{q}$ is a rational root (expressed in reduced form) of the polynomial with integer coefficients $a_0 + a_1 x + a_2 x^2 + \cdots + a_n x^n$, then p is a factor of the constant term a_0, and q is a factor of the leading coefficient a_n.

This limits the search for any rational roots to the testing of a finite set of possible candidates. In particular we may completely solve any cubic with rational coefficients that has some rational solution as follows. First clear the denominators to get an equation with integer coefficients, and find a rational root r by the above method. Then factorise the polynomial in the form $(x - r)q(x)$, where $q(x)$ is a quadratic that we may solve in the usual way, extracting the final two roots, even if they should happen to be complex numbers.

The theorem is true as a consequence of elementary factoring properties of integers. Substituting our root into the polynomial gives:

$$a_0 + a_1 \frac{p}{q} + \cdots + a_n \frac{p^n}{q^n} = 0 \Rightarrow a_0 q^n + a_1 p q^{n-1} + \cdots + a_n p^n = 0$$

Then since p and q have no common factor, it follows that q must be a factor of a_n, as q is a factor of every other term in this equation. Similarly p is a factor of a_0, as p is a factor of all the other terms.

The theorem is really quite powerful, as we may deduce immediately that the nth root of an integer k is either an integer or is irrational (see Note 27) by considering the polynomial $x^n - k$;

by the Rational Root Theorem, for any rational root $\frac{p}{q}$ of this polynomial, q must be ± 1, from which the conclusion follows.

Chapter 10

Note 42 Page 186 footnote, *Quarter Square Rule*

The purpose of the logarithm table was to replace difficult multiplications with relatively easy additions. However, a very simple trick for expressing products in terms of sums found in some old table books is the *Quarter Square Rule*, which is nothing more than the easily verified identity:

$$ab = \frac{1}{4}(a + b)^2 - \frac{1}{4}(a - b)^2.$$

Armed with a table of quarter squares, one can therefore compute any product as the difference of the two quarter squares that arise from the sum and difference of the two numbers to be multiplied. For example, to calculate 228×139, we put $a = 228$ and $b = 139$ and look up the quarter squares of $a + b = 367$ and $a - b = 89$. The table will give the two values 33,672.25 and 1980.25, and we take their difference to obtain $228 \times 139 = 31,692$. (Actually the fractional parts, .25 in this case, will always be the same, and so will cancel and so can be ignored—only the integer part need be listed in the table.) What is more, the answer is exact, unlike the outcome through use of a log table or slide rule, which is always an approximation. In principle, the technique is not confined to multiplication of integers but applies to any numbers whose quarter squares are tabulated. Indeed suitable scaling will allow for non-integral values, for instance this example in effect shows that $2.28 \times 13.9 = 3\ 1692$.

Note 43 Page 193 footnote, & 201 *Multiplication and Division of Complex Numbers*

Assuming the Distributive Law continues to hold, the multiplication in rectangular form proceeds as follows:

$$zw = (a + bi)(c + di) = a(c + di) + bi(c + di)$$
$$= ac + adi + bci + bdi^2 = (ac - bd) + i(ad + bc)$$

which is Hamilton's expression.

Division on the other hand can be calculated directly by means of the *complex conjugate*. In general, the conjugate of $z = a + bi$ is denoted by $\bar{z}$ and is $a - bi$, in other words $\bar{z}$ is the reflection of z in the real axis. By the multiplication rule we see that $z\bar{z} = a^2 + b^2$. This is a real number, and equals the square of the distance of z from the origin, denoted by $|z|$. In symbols $z\bar{z} = |z|^2$. We may now divide one complex number by another by multiplying top and bottom by the conjugate of the divisor, in order to make the division one by a purely real number. This is analogous to the familiar rationalising of the denominator that is used to remove the surd in a division involving square roots. In detail we have

$$\frac{z}{w} = \frac{a + bi}{c + di} = \frac{(a + bi)(c - di)}{(c + di)(c - di)} = \frac{((ac + bd) + i(bc - ad))}{c^2 + d^2}$$
$$= \frac{ac + bd}{c^2 + d^2} + i\frac{bc - ad}{c^2 + d^2}$$

Of course, as with adding fractions, as long as the technique is understood, there is no need to memorize the answer.

Note 44 Page 196 & 200 *Polar Form and De Moivre's Theorem*

The rectangular coordinates of $z = (r, \theta)$ are given by $z = r \cos \theta + i r \sin \theta$. Hence if $z_1 = (r_1, \theta_1)$ and $z_2 = (r_2, \theta_2)$ are two complex numbers in polar form, using the technique of the previous note and the addition formulae for cosine and sine we obtain

$$z_1 z_2 = r_1 r_2 (\cos \theta_1 \cos \theta_2 - \sin \theta_1 \sin \theta_2)$$
$$+ i r_1 r_2 (\cos \theta_1 \sin \theta_2 + \cos \theta_2 \sin \theta_1)$$
$$= r_1 r_2 (\cos(\theta_1 + \theta_2) + i \sin(\theta_1 + \theta_2))$$

which, returning to polars gives $z_1 z_2 = (r_1 r_2, \theta_1 + \theta_2)$.
Applying this repeatedly to positive powers then gives $z^n = (r, \theta)^n = (r^n, n\theta)$. Since it can be checked directly that $z^{-1} = (r^{-1}, -\theta)$ it follows that this formula, known as De Moivre's Theorem, holds for positive and negative powers alike, and for fractional powers as well.

Note 45 Page 202 *The hyperbolic functions*

In general, a function $f(x)$ may be written uniquely as a sum $e(x) + o(x)$ of an *even function* $e(x)$, and an *odd function* $o(x)$, meaning that $e(x) = e(-x)$ and $o(-x) = -o(x)$ for all values x. Even and odd functions are respectively characterised by the properties that the graph of an even function is symmetric with respect to reflection in the y-axis, and, in the odd case, the graph is the same when rotated through $180°$ about the origin. For example, x^2 and $\cos x$ define even functions, whereas x^3 and $\sin x$ are odd. The even and odd parts of $f(x)$ are easily verified to be

$$e(x) = \frac{f(x) + f(-x)}{2}, \; o(x) = \frac{f(x) - f(-x)}{2}$$

Applying this to the exponential function $f(x) = e^x$, we get the even and odd parts that are called respectively *hyperbolic cosine* and *hyperbolic sine:*

$$\cosh(x) = \frac{e^x + e^{-x}}{2}, \; \sinh(x) = \frac{e^x - e^{-x}}{2}$$

Note 46 Page 202 footnote *Osborne's Rule*

For every trigonometric identity there is a corresponding hyperbolic identity that is easily verified from the definition in each individual case. For example

$$\cos^2 x + \sin^2 x = 1; \cosh^2 x - \sinh^2 x = 1$$

$$\sin 2x = 2 \sin x \cos x; \sinh 2x = 2 \sinh x \cosh x$$

This relation type is explained by Euler's formula $e^{i\theta} = \cos\theta + i\sin\theta$, for accepting this we obtain at once

$$-i \sinh(ix) = -i\frac{e^{ix} - e^{-ix}}{2} = \frac{-i}{2}((\cos x + i\sin x)$$
$$- (\cos(-x) + i\sin-(x))) = \sin x$$
$$\cosh ix = \frac{e^{ix} + e^{-ix}}{2} = \frac{1}{2}((\cos x + i\sin x)$$
$$+ (\cos x - i\sin x)) = \cos x$$

Or equivalently

$$\sin(ix) = i\sinh x \text{ and } \cos(ix) = \cosh x.$$

Meaning can be attached to imaginary arguments of cosine and sine through use of series expansions of these functions.

For example, replacing x by ix in the Pythagorean identity now gives

$$1 = \cos^2 ix + \sin^2 ix = \cosh^2 x + i^2 \sinh^2 x = \cosh^2 x - \sinh^2 x.$$

$$\sin(2ix) = 2\sin(ix)\cos(ix) \Rightarrow i\sinh(2x) = 2i\sinh x \cosh x$$

$$\Rightarrow \sinh 2x = 2\sinh x \cosh x.$$

Osborne's Rule encapsulates the transformation from the trigonometric to the hyperbolic identity:

Replace each trigonometric function by its hyperbolic counterpart, and *change the sign of each term that involves the product of two hyperbolic sines.*

This accounts for the change of sign that we see in the first example, which did not appear in the second.

Note 47 Page 205 footnote *Sums and differences of Squares*

Let S_i denote the set of integers that are the sum of i squares. The nice thing about S_1, S_2 and S_4 is that they are all semigroups under multiplication. This can be interpreted as following from the multiplicative property of the norm (or modulus) for the reals, the complex, and the quaternion numbers respectively. The trouble with S_3 is that it is not a semigroup—for example $3 = 1^2 + 1^2 + 1^2$ and $13 = 0^2 + 2^2 + 3^2$ yet the product, $3 \times 13 = 39$, is not the sum of three squares. This is part the reason why the theorem of Gauss that a number is a sum of three squares if and only if it does *not* have the form $4^e(8k + 7)$ is difficult to prove, at least in one direction. It is however quite easy to show that no number of the proscribed form is the sum of three squares.

First observe that since a square is congruent modulo 8 to one of 0, 1 or 4 and that 7 cannot result from summing three numbers from this set (even allowing for repeats) it follows that no number of the form $8k + 7$ is in S_3. Next we show that if an integer d is the sum of three squares and that $4|d$, then $\frac{d}{4}$ is also a member of S_3. Given that this claim is valid, it follows that d cannot have the form $4^e(8k + 7)$ for then applying the claim e times would yield the false conclusion that a number of the form $8k + 7$ was the sum of three squares.

To prove the claim let us assume that $d = 4m$ and that $d = a^2 + b^2 + c^2$. If one or more of the three squares were congruent to 1 modulo 8 then we would get that $d \equiv 1, 2, 3, 5,$ or 6 modulo 8 but since $4|d$ we have that $d \equiv 0$ or $d \equiv 4 \pmod 8$ and so this is not possible. Hence each of the squares is congruent to either 0 or 4 modulo 8 and in particular each square is divisible by 4 and so each of a, b and c are even. However, dividing through by 4 then gives immediately that $m = (a/2)^2 + (b/2)^2 + (c/2)^2$, as required to establish the claim and so complete the proof that no number of the form $4^e(8k + 7)$ is the sum of three squares.

It is easy to see that an integer n is the *difference* of two squares if and only if $n \not\equiv 2 \pmod 4$. This follows from the observations that any odd number $2k + 1 = (k + 1)^2 - k^2$, while $4k = (k + 1)^2 - (k - 1)^2$. However, for any difference of two squares, $a^2 - b^2 = (a - b)(a + b)$; since the two factors $a - b$ and $a + b$ differ by an even number, these factors are either both even, giving a number of the form $4k$, or both odd in which case the product is odd also. Hence no number of the form $4k + 2$ is the difference of two squares.

Chapter 11

Note 48 Page 211 *Farey Fractions & Euler totient function*

Apart from F_1, each Farey sequence F_n has $\frac{1}{2}$ in the middle, with an equal number of terms placed symmetrically either side. In particular the total number of terms is odd. The exact number of terms $N(n)$ is $1 + \sum_{k=1}^{n} \phi(k)$, where $\phi(k)$ is Euler's *totient function*, the number of numbers less than k that share no common factor with k. (We say such a pair of numbers is relatively prime.) The totient function is multiplicative, meaning that if m and n are relatively prime, then $\phi(mn) = \phi(m)\phi(n)$. It follows that a formula for $\phi(k)$ can be given in terms of its prime decomposition, as for a prime power p^t, it is easy to see that $\phi(p^t) = p^{t-1}(p - 1)$. Indeed $\phi(k)$ can be found as long as the prime divisors of k are known, for it follows readily from all this that:

$$\phi(k) = k \left(1 - \frac{1}{p_1}\right) \left(1 - \frac{1}{p_2}\right) \cdots \left(1 - \frac{1}{p_r}\right),$$

where the p_i are the distinct prime factors of k.

The first few values for the sequence of lengths $N(n)$ are then seen to be $2, 3, 5, 7, 11, 13, 19, \cdots$ and it is known that, in the limit, the value of $N(n)$ approaches in ratio $\frac{3n^2}{\pi^2}$.

Note 49 Page 221 *Euclid's Lemma*

Suppose that p is prime and a factor of ab $(1 < a, b < p)$ so that $ab = rp$ say. Either p is a factor of a (whence we are finished) or not. In the latter case, *since p is prime*, the hcf of a and p is 1. By the euclidean algorithm we may write 1 in the form $1 = ax + py$

for some integers x and y. Now

$$b = b \times 1 = b(ax + py) = bax + bpy$$

Since $ba = pr$, we substitute in this last equation accordingly to obtain

$$b = prx + pby = p(rx + by)$$

But this shows that p is a factor of b, as required to complete the proof of Euclid's Lemma.

Note 50 Page 222 *Continued fraction representation of $\sqrt{2}$.*

There are two steps in the calculation of a continued fraction for a number $x = [a_0, a_1, a_2, \ldots]$. The number a_0 is the integer part of x, denoted by $a_0 = \lfloor x \rfloor$. In general $a_n = \lfloor r_n \rfloor$, the integer part of r_n, where the remainder term r_n is defined recursively by $r_0 = x$, $r_n = \frac{1}{r_{n-1} - a_{n-1}}$. Applying this to $x = \sqrt{2}$ we get:

$$x = r_0 = \sqrt{2} = 1 + (\sqrt{2} - 1) \text{ so that } a_0 = 1$$

and

$$r_1 = \frac{1}{\sqrt{2} - 1} = \frac{\sqrt{2} + 1}{(\sqrt{2} - 1)(\sqrt{2} + 1)} = \sqrt{2} + 1, a_1 = \lfloor r_1 \rfloor = 2;$$

Next we get

$$r_2 = \frac{1}{r_1 - a_1} = \frac{1}{(\sqrt{2} + 1) - 2} = \frac{1}{\sqrt{2} - 1}$$
$$= \frac{\sqrt{2} + 1}{(\sqrt{2} - 1)(\sqrt{2} + 1)} = \sqrt{2} + 1,$$

so that

$$r_1 = r_2 = \cdots, a_1 = a_2 = \cdots = 2; \text{ and so } \sqrt{2} = [1, \overline{2}].$$

In general a continued fraction $[a_0, a_1, a_2, \cdots]$ can also be represented as $a_0 + (1 + (a_1 + (1 + a_2(1 + \cdots)^{-1})^{-1})^{-1} \cdots)^{-1}$.

Note 51 Page 224 *Continued fraction and representations involving e*

The standard series for e as the sum of the reciprocals of the factorials leads to a representation as the following nested product:

$$e = 1 + 1 + \frac{1}{2}\left(1 + \frac{1}{3}\left(1 + \frac{1}{4}\left(1 + \frac{1}{5}\left(1 + \cdots\right)\right)\right)\right),$$

The *simple* continued fraction representation of e, that is the one using only unit denominators is:

$$e = [2, 1, 2, 11, 4, 11, 6, \cdots].$$

Other simple continued fractions involving e are:

$$e - 1 = [1, 1, 2, 1, 1, 4, 1, 1, 6, \cdots] \text{ and}$$

$$\frac{1}{2}(e - 1) = [0, 1, 6, 10, 14, \cdots].$$

These and further examples can be found on the web page http://mathworld.wolfram.com/e.html.

Note 52 Page 227 *ternary fraction to decimal*

Work as in Note 24 but this time in ternary:

$$a = \overline{0 \cdot 20}_3 \Rightarrow 100_3 a - a = 22_3 a = 20_3$$

$$\Rightarrow a = \left(\frac{20}{22}\right)_3 = \left(\frac{10}{11}\right)_3 = \frac{3}{4}.$$

Chapter 12

Note 53 Page 231 *Caesar ciphers*

More effective substitution ciphers can be created using a little modular arithmetic. Number the letters of the alphabet from 0 to 25, so that A is 0, B is 1 and so on down to Z is 25. A shift cipher then corresponds to the rule $a \to a + b$ say, where b is the shift in the alphabet. A simple linear shift such as $a \to 3a + 2$ is harder to decipher. In this example $A \to C$, $B \to E$, $C \to H \cdots$, $F = 5 \to 17 = R, \cdots M = 12 \to 38 \equiv 12 = M, \cdots, Z = 25 \to 77 \equiv 25 = Z$. The pattern of the substitution then appears much more random. The cipher is still vulnerable however to basic frequency analysis. In general a linear substitution cipher $a \to ka + b$ (mod 26) will only be one-to-one if k is relatively prime to 26. For example $a \to 2a + 3$ would lead to ambiguity as $A \to D$ and $N \to D$ also.

Note 54 Page 250 *Remainders of powers*

What is being invoked here is a special case of the facts, easily verified from the definition, that if $a \equiv a'$ and $b \equiv b'$ both mod m, then $ab \equiv a'b'$ (mod m). It follows from this, by taking $a = a'$ and $b = b'$ and using induction on n that if $a \equiv b$ (mod m) then $a^n \equiv b^n$ (mod m) for all $n = 1, 2, \cdots$. Now any number t is congruent mod m with its remainder r when divided by m. Hence if $2^a \equiv r$ (mod m) with $0 \leq r \leq m - 1$, then we have $2^{ab} = (2^a)^b \equiv r^b$ (mod m), which is the number calculated by Alice. Similarly the number calculated by Bob is congruent to $2^{ba} = 2^{ab}$ (mod m). Since both Alice and Bob's final numbers are in the range 0 to $m - 1$, they are equal to one another.

Note 55 Page 256 *Fast exponentiation*

Again the trick, as introduced in the previous note, is that, when doing modular arithmetic we may replace any number by another that is congruent to it modulo m. The other computational idea to exploit is to write the index as a sum of powers of 2: in this case $15 = 8 + 4 + 2 + 1$, because successive powers of 2 modulo m are readily realized. Working modulo 1081 we find that $77^2 = 5929 \equiv 5 \times 1081 + 524 \equiv 524 \pmod{1081}$; $77^4 \equiv 524^2 = 274,576 = 254 \times 1081 + 2 \equiv 2 \pmod{1081}$; $77^8 \equiv 2^2 = 4 \pmod{1081}$. Hence

$$77^{15} \equiv 4 \times 2 \times 524 \times 77 = 616 \times 524 = 308 \times 1048$$
$$\equiv 308 \times (-33)$$
$$= 924 \times (-11) \equiv (-157) \times (-11) = 1727$$
$$\equiv 646 \pmod{1081}.$$

Note 56 Page 257 *Calculation of Alice's decoding number d*

(See Note 48 for explanation of Euler's totient function.) Using the formula for Euler's function we see that $\phi(n) = \phi(pq) = (p - 1)(q - 1)$. However this can be calculated directly: the multiples of p less than n are: $p, 2p, 3p, \cdots, (q-1)p, qp$, which number q in all, while those for q similarly number p altogether. They have one common multiple, that being pq itself, so that $\phi(n) = pq - p - q + 1 = (p - 1)(q - 1)$. Alice needs to find a value of d such that $ed \equiv 1 \pmod{\phi(n)}$. The reason for this is as follows. Bob has sent her the number encrypted as $M^e \pmod{n}$. A fundamental property of the Euler ϕ function is that is satisfies $a^{\phi(n)} \equiv 1 \pmod{n}$, for any $1 \le a \le n - 1$. To say that d satisfies the above equation means just that $ed = 1 + k\phi(n)$ for some integer k. But then we have:

$$M^{ed} = M^{(1+k\phi(n))} = M \cdot (M^{\phi(n)})^k \equiv M \cdot 1^k = M(\text{mod } n);$$

so that Alice can recover the message M by doing this sum. The existence of d is guaranteed by the fact that e and $\phi(n)$ are chosen to be relatively prime so that, by Euclid's algorithm, there exist integers x and y such that $ex + \phi(n)y = 1$; hence working modulo $\phi(n)$ we see that $ex \equiv 1 \pmod{\phi(n)}$. If $x < 0$ we may add sufficient multiples of $\phi(n)$ to x to find a positive number d that will also satisfy $ed \equiv 1 \pmod{\phi(n)}$, as Alice requires.

Note 57 Page 258 *Bob and Alice's calculations*

Bob: $6^2 = 36 \equiv 3 \pmod{33}$; $6^4 \equiv 3^2 = 9 \pmod{33}$ so that $M^e = 6^7 = 6^4 \times 6^2 \times 6 \equiv 9 \times 3 \times 6 \equiv 27 \times 6 \equiv (-6) \times 6 \equiv -36 \equiv -3 \equiv 30 \pmod{33}$.

Alice: $M^{ed} = 30^3 \equiv (-3)^3 = -27 \equiv 6 \pmod{33}$, so that Alice has Bob's plaintext message, $M = 6$.

Further Reading

A real insight into the nature of numbers can be read in David Flannery's new book, *The Square Root of 2: A Dialogue Concerning a Number and a Sequence* (Copernicus Books, 2006). The entire book is in the Socratic mode of a conversation between a teacher and pupil. Like any written dialogue, it is ultimately artificial, but very much to a purpose. The nature of the irrational is explored very thoroughly. A formal mathematics text could 'cover' the same material in fewer pages but the full force of the content would not be so keenly felt, even by a well-trained mathematics student, than when it is developed in the patient and natural fashion of this book.

If you are ready for a real mathematical style introductory text you can't do better than *Elementary Number Theory* by G. Jones & J. Jones (Springer-Verlag, 1998). It gives a reasonably gentle but rigorous introduction and goes as far as aspects of the famous Riemann Zeta Function and Fermat's last theorem. An old book that I particular like is Underwood Dudley's book of the same title which is especially gentle, but again rigorous and tackles some tougher topics toward the end. Be warned, you can't judge a book by its title. For instance Andre Weil's classic text *Basic Number Theory* begins with the claim that 'no knowldege of number theory is pre-supposed' but then lists a series of mathematical pre-requisites to be grasped before wading into the text including 'the existence and unicity of the Haar measure'. A serious mathematical text that does start from scratch however is perhaps the most famous book on numbers, *An Introduction to the Theory of Numbers,* by G.H. Hardy and G.M. Wright (Oxford, OUP) which is still in print after seventy years. Although it assumes little particular mathematical knowledge, it hits the ground running!

A popular account of the Riemann zeta function is the book by Carl Sabbagh, *Dr Riemann's Zeros* (Atlantic books, 2003). Unlike Fermat's Last Theorem, which can be explained to anyone in a couple of minutes, the Riemann conjecture (see Note 1 of Chapter 13 above) is much more technical so it is a real challenge to engage the general reader with this, perhaps the biggest remaining open question in mathematics. He does a good job, as does Marcus du Sautoy in his *The Music of the Primes, Why an Unsolved Problem in Mathematics Matters* (Harper Collins, 2004), which is essentially on the same topic. In his offering, du Sautoy engages the subject a little more directly, but it still written for the benefit of the general reader.

There are two very good and very different accounts of the solution to Fermat's last Theorem, those being *Fermat's Last Theorem: Unlocking the Secret of an Ancient Mathematical Problem* by Amir D. Aczel (Penguin, 1996) and *Fermat's Last Theorem* by Simon Singh (London, Fourth Estate, 1999). *Fermat's Last Theorem for Amateurs* by Paulo Ribenboim (Springer-Verlag, 1999) on the other hand is an account of the mathematics surrounding the problem. The best popular book about on the history of coding up to the RSA cipher is also an effort of Simon Singh: *The Code Book* (Fourth Estate, 2000).

The Book of Numbers by John Conway and Richard Guy (New York, Springer-Verlag, 1996) is full of history, vivid pictures, and all manner of facts about numbers. It is not a textbook but nonetheless the authors are keen to explain as completely as possible everything they find of interest on the topic. Paul Halmos's *Naive Set Theory* (New York, Springer-Verlag 1974) gives a quick introduction to infinite cardinal and ordinal numbers. It is a mathematics text, and some experts say it is a little dated now as the subject of Set Theory has moved on. However the book is short enough to be able to read right through and will set the reader up with the right reflexes as regards the way sets in general and infinite numbers in particular behave.

The insolvability of the quintic (fifth degree polynomial equations) was only touched on the text here but a truly interesting book for historians of the subject is *Abel's Proof: An Essay on the Sources and Meaning of Mathematical Unsolvability* (MIT Press, 2003) by Peter Pesic. Abel's original method was superseded by Galois but Pesic went back and learnt exactly how Abel originally proved this most famous of negative results: you cannot solve 5th degree equations the way you can for those of lower degree. It is quite accessible for the mathematically able and rather refreshing in its direct approach.

There are number theory novels about. Two of them are *The Parrot's Theorem* (London, Orion fiction, 2000) by Denis Guedj, a mystery dedicated ot Fermat's theorem and the Goldbach conjecture while another entertaining effort is *The Wild Numbers* by

Philibert Schogt (London, Orion fiction 2000) which captures the triumphs and delusions of real research mathematics in a way that most readers will find surprising.

Two general histories of mathematics that are excellent sources are *A History of Mathematics* by Carl. B. Boyer (New York, Wiley, 1968) and *An Introduction to the History of Mathematics* by Howard Eves, (New York, Holt, Reinhart and Winston, 1969). For a more biographical approach, E.T. Bell's *Men of Mathematics* (some of whom are women) is always popular (New York, Simon and Schuster, 1937). A more modern text is John Stillwell's *Mathematics and its History* (New York, Springer-Verlag, 1991 and 2002) which is an excellent and unusual book in that it teaches mathematics but in its historical context. *An Imaginary Tale: The Story of $\sqrt{-1}$* (Princeton University Press, 1998) presents quite a lot of detail on the history surrounding complex numbers but this book primarily celebrates the magic of the number system itself, told from the viewpoint of the author, Paul J. Nahin, a professor of electrical engineering.

A very high quality web page that allows you to dip into any mathematical topic, and is especially rich in number matters, is Eric Wolfram's MathWorld: mathworld.wolfram.com. For mathematical history topics I particularly recommend *The MacTutor History of Mathematics archive* at St Andrews University, Scotland: www-history.mcs.st-andrews.ac.uk/history.index.html.

Index

A

Abel, N. 183, 317
addition 24, 108
addition principle 18
Akhmim papyrus 107, 285–86
algebraic identity 32
Al-Kashi 110, 128, 181
amp 99
Archimedes 21, 117
Argand, J. 190
arithmetic progression 69, 95
arithmetic triangle 63–64, 140, 271–72
Ars Magna 175
average 90–97
axis
   real 193
   imaginary 193

B

Babbage, C. 235
Babylonians 5
ballot problem 151–52
base: 5–8
   hexadecimal 5
   sixty 5, 22, 110
   ten 5
   three 112
   twelve 5
   twenty 5
   two 5, 30, 292
Bernoulli, D. 157
Bertrand, J. 162, 202
   paradox 161–62
   postulate 79
binary (*see base two*)
binary code 240

binomial coefficients 63–65
Binet's formula 274
birthday problem 152, 297
Bombelli, R. 181, 188
Briggs, H. 90, 186
Buffon, Comte de 158
   needle problem 160–62
Buffon-Laplace problem 160–61

C

Cantor, G. 122, 228
   diagonal argument 123
   middle third set 132, 225–28
cardinal (*see number*)
Cardano, G. 139, 175–77
casting out nines 35–39
Cauchy, A. 191, 203
Chu Shih-chieh 64, 182
Chuquet, N. 172, 187
cipher: 229
   book 240–41
   Caesar 230, 312
   one-time pad 239
   RSA 254
   substitution 232
   Vigenère 233–37
ciphertext 230
circuit:
   parallel 98–100
   series 99–100
Cocks, C. 252
coin problem 217
Collatz algorithm 76
complex (*see number*)
combinatorics 102–3
compasses 113

completing the square 173, 300
complex conjugate 181, 199,
        304
convergents 223
coordinates
    rectangular 190–94
    polar 195–98
cot deaths 147
cryptography 229–55

D

decimal: 109–10
    recurring 110, 287–88
    terminating 111
Dedekind, J. 191
Delian problems 115, 118
del Ferro, S. 175
De Moivre A. 200
    Theorem 200, 305–6
denominator 107
    common 108
    lowest common 108
derangement 294
Descartes, R. 187
difference of squares 307
Diffie, W. 246
digital signature 258
Diophantus 20
Dirichlet, J. 79
dividend 28
divisibility tests 39–45, 268–69
division 27, 107–8
divisor 28
domino 34
duodecimal 41
Durer, A. 53

E

e 82, 126–27, 292–95, 311
equation:
    Barker's 182
    cubic 174–83, 302

diophantine 178
    linear 168
    quadratic 172, 179–81
    quartic 175
    quintic 182
Erdös, P. 76
Euclid 14
euclidean algorithm 214–15, 220, 309,
        314
Euclid's Lemma 221, 309–10
Euler, L. 61, 75, 189
    formula 306
    totient function 255, 309
even function 202, 305
event space 138
exponentiation 26

F

factor 108
factorial 12, 127
Farey sequence 210–12, 309–10
fast exponentiation 313
Ferrari, L. 175
Fermat, P. 61, 187
Fermat's Lemma 13, 83, 280–81
ferry boat problem 170–71
Fibonacci 67, 221
Fior, A. 175
Fourier transform 204
fractions: 105–15, 210
    continued 215, 310–11
    egyptian 106–08, 212, 284
    reduced 212
    unit 106, 212
frequency analysis 232
Fundamental Theorem of Arithmetic 9,
        199, 221, 289

G

Galileo, G. 2, 4, 119
Galois, E. 183
gamma function 264

Gauss, K. 82, 191
geometric progression 69, 95
GIMPS 15
Girard, A. 188
global warming 169
Goldbach's Conjecture 80
golden ratio 70–71, 212, 221, 274
golden rectangle 71, 274
groups: 15
  monster 15
  simple 15

H
Hamilton, W. R. 194, 204–5
Hardy, G.H. 13
harmonic progression 96–97
Hellman, M. 246
Heron of Alexandria 97
  iteration 282
highest common factor 108, 212–13
Hilbert, D. 120, 167
  hotel 120–22
hyperbolic function 202
  cosine 306
  sine 306

I
i 181, 188
irrational (*see number*)
imaginary (*see number*)
incommensurable 114
integer: 104, 123
  gaussian 198–200
inverse proportion 99

K
Kasiski, F. 236
Khayyam, Omar 168, 174
Kepler, J. 68
key exchange 243–54

L
Laplace, P-S. 139
Leibniz, G. 188
Lichtenburg ratio 71, 275
Lindemann, F. 119
logarithm: 189, 295
  common 90, 186
  natural 126–27
lo shu 48
Loyd, S. 170

M
magic circle 58
magic hexagon 59
magic pentagram 57
magic square: 48–52
  complementary 52, 270
  normal 48
matrix 206
Mauborgne, J. 238
mean: 90–97, 281–83
  arithmetic 90–92
  common 90–92
  geometric 96–97
  harmonic 96–97
measure theory 132
median 93–94
mode 94
modular arithmetic 249, 255, 312
modulus 38, 196
multiplicands 36
multiplication 25
multiplicative function 266, 309

N
Napier, J. 90, 185
neighborhood 228
Newton, I. 188, 195
Nicholas of Cusa 4
norm 205, 307

number:
abundant 13
Ackermann 24
algebraic 123, 289–90
amicable 61, 271
Bell 74–75
cardinal 124–25, 291
Carmichael 14
Catalan 65–67
choice 11
complex 103, 190–208, 304–7
composite 9, 275
decoding 256, 313
encoding 256, 313
euclidean 115, 123
Fermat 267, 278–79
Fibonacci 8, 67–72, 221–22, 272, 274–75
Frobenius 219
hailstone 76–77
imaginary 103, 181, 190
irrational 113, 209, 295–96
Lucas 68
lucky 84–85
Mersenne 277
natural 101
negative 104
ordinal 134
partition 75
perfect 11, 277
primal 9
prime 9 (see primes)
rational 112, 123, 209, 295–96
real 115, 119, 123, 190
rectangular 9
sociable 62
Stirling 72–73, 273
surreal 136
transcendental 123, 212, 290–91
triangular 11
number line 104, 128, 209–215, 227–28
numerator 107

O
octonians 206
odd function 202, 305
odds 86, 127
ohm 99
origin 193
ordinal (see number)
Osborne's Rule 306
outlier 92

P
palindrome 48, 233
partial fractions 103
partial sum 133
Pascal, B. 139
Pascal's triangle (see arithmetic triangle)
percentage 85–86
percentile 86, 93
pi 113, 117, 128
pigeonhole principle 111, 286–87, 297
plaintext 230
polar:
    coordinates 195
    form 305–6
pole 195
Poisson, S. 216
Poisson process 155–56
Poulet P. 62
positional principle 7
primes: 9, 77–83, 254, 263, 275–78
    Mersenne 15, 266, 277
    twin 80
prime factorization 9
probability: 138–58
    conditional 148
public key encryption 252
Pythagoras: 1, 113–14
    Theorem 1, 165

Q
quadratic (see equation)
    formula 300–1

quadrivium 97
Quarter Square Rule 186, 303
quartile 86
quaternions 205, 307
quotient 28

R
rabbit problem 67–68
radian 197
Ramanujan, S. 13, 75, 187
ratio 86–88
rational (see number)
Rational Root Theorem 302–3
real line (see number line)
recurrence relation 73
Regiomontanus 166
relatively prime 215
remainder 28
resistor 98–99
Riemann Conjecture 9, 263
Riemann zeta function 263
Roman numerals 18
root 289
Russell, B. 191, 203
Russian roulette 152–53

S
scientific notation (see standard form)
set: 119
    countable 123
    dense 130
    empty 125
    finite 125
    infinite 125
    middle third 225–28
    nowhere dense 227
    sub- 119
    uncountable 123

sieve of Eratosthenes 78
Siam Method 54
squaring the circle 118, 201
standard form 88–89
St Petersburg paradox 157–58
Stevin, S. 110
straightedge 113
Stifel, M. 174
Stirling's triangle 73–74, 273
subtraction 24, 108
subtraction principle 22
sums of squares: 307–8
    four 200, 205
    three 205, 307–8
    two 199–201
sunya 7
Syracuse problem 76

T
Tartaglia, N. 175–77
ternary 112, 226, 311
torus 54
trapdoor function 252
trivium 97
Turing, A. 242

V
vector 193
Viete, F. 110, 186
    substitution 301–2
voltage 98

W
Wallace, J. 191
    product 223
Well, R. 221
Wessel, C. 190
Wilson's Theorem 83

Printed in the United States
By Bookmasters